Fundamentals of Office Ergonomics

Edited by Peter Budnick, Ph.D., CPE

An Ergoweb® Inc. Publication

www.ergoweb.com

ISBN 0-9749843-0-2

Acknowledgements

This book is a compilation of work that has evolved over many years through the hard work of various direct and indirect contributors. Among those who deserve credit and thanks – and this is by no means a complete list – are Tim Villnave, Rachel Michael, Jeanie Croasmun, Tim Cherry, Ryan Jeppesen, and Don Bloswick. Additional contributors are identified with their contributions throughout this book.

Thanks also goes to Ergoweb Inc. and it's employees for supporting this work and the various services that assisted in its development, including the news services *The Ergonomics Report*™ and *Ergonomics Today*™, *ergoweb-L*™, the world's largest ergonomics specific discussion forum, and the *Ergonomics Image Gallery*™, the source of many of the illustrations used throughout this book.

And finally, Ergoweb Inc. wouldn't exist without the loyal and enthusiastic support of our customers. Thank you all for your continued support.

Peter Budnick

Table of Contents

Selected Articles

1. Productivity, Quality, Health and Safety: A Strong Business Case for Ergonomics

1.1 Introduction

Ergonomics is a scientific and applied discipline that seeks to improve human performance.

Over the past decade or so, the term "ergonomic" has entered popular culture as a marketing buzzword to describe chairs, dashboards, electronics, tools, fitness equipment and even salsa chips. This use of the word gives an intuitive sense as to what ergonomics is about: designing something to fit well or work easily.

Wojciech Jastrzebowski

Wojciech Jastrzebowski, a Polish scholar, first used the term ergonomics in 1857 in a philosophical narrative "based upon the truths drawn from the Science of Nature" (Jastrzebowski, 1857). He derived it from the Greek word *ergon* (work) and *nomos* (principle or law) to mean the Science of Work. Jastrzebowski was careful to note that his use of the word "work" had very broad implications, not just in the occupational sense:

> *[T]his Science of Work, understood as Work in the comprehensive and integral sense, not merely its part that is physical labor or toil, but physical, aesthetic, rational, and moral work, that is Labor, Entertainment, Reasoning, and Dedication ...*

Further, he recognized that good ergonomics enriched life outside the workplace.

> *... [W]e should live to make our lives as bountiful as possible at the least expenditure of toil and with the greatest inner contentment, bringing forth fruit for our own and for the common good.*

1.2 Ergonomics, Productivity and Quality

Ergonomics has deep roots in the productivity improvements that characterized much of the technology advancements of the 1900s. Fredric Taylor, for example, achieved dramatic productivity improvements in the steel production industry by closely studying the work and discovering optimal relationships between specific tools and tasks, and the people that conducted the work. He was able to maximize the amount of material handled in a day, reducing wasted effort and increasing employee compensation in the process. In his most famous example, he matched the size and shape of shovels with the type of material being shoveled to optimize production.

Henry Ford, of automobile fame, also made dramatic improvements in productivity by designing production lines that organized the work into an efficient process. He is credited with saying (paraphrasing) "bring the work to the worker, don't make the worker go to the work." In essence, he was saying minimize unnecessary motions and reaching by designing the work to match the worker's most comfortable and productive positions and postures.

Frank and Lillian Gilbreth are also notable innovators in productivity improvements through the development of micromotion studies, which have evolved into the motion-time studies used today. By studying micromotions in great detail, the Gilbreths were able to assign reliable time estimates to each type of task (e.g., reach, grasp, move, release). This provided a framework in which to define and monitor productivity as it relates to human task motions.

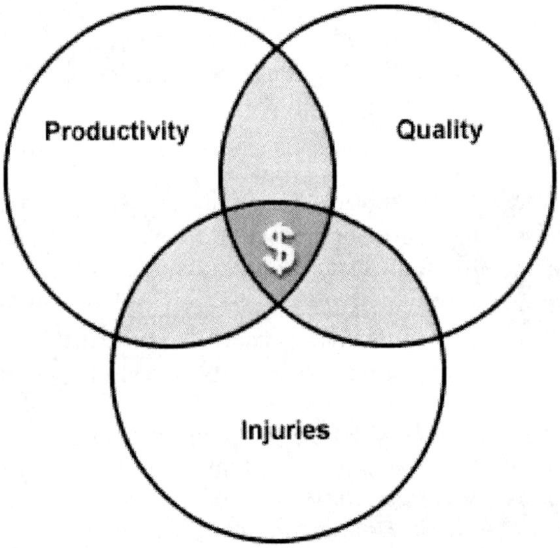

Reprinted from the *Ergonomics Image Gallery*™ with permission from Ergoweb Inc. Copyright © 2004.

Body motions, visibility, workload and other important ergonomic parameters will also affect the quality of work and the quality of work product. When a task is matched with the ability of the people that will perform it, they will make fewer errors and produce less waste. Ergonomic design considerations have also been shown to influence employee recruitment and retention.

Any ergonomics intervention must be viewed in light of its effect on productivity and quality, and the best ergonomics solution will often improve both metrics while reducing injury potential and associated costs. Simply put, reducing unnecessary or awkward postures, motions and exertions almost necessarily reduces the time and energy it takes to complete a given task.

1.3 Ergonomics, Health and Safety

The association between occupations and musculoskeletal injury was also documented early. Bernardino Ramazzini (1633-1714) wrote about work related complaints that he saw in his medical practice in the 1713 supplement to his 1700 publication *De Morbis Artificum (Diseases of Workers)*:

The maladies that afflict the clerks aforesaid arise from three causes: First, constant sitting; secondly, the incessant movement of the hand and always in the same direction, thirdly, the strain on the mind from the effort not to disfigure the books by errors or cause loss to their employers when they add, subtract, or do other sums in arithmetic.... In a word, they lack the benefits of moderate exercise, for even if they wanted to take exercise, they have no time for it; they are working for wages and must stick to their writing all day long. Furthermore, incessant driving of the pen over paper causes intense fatigue of the hand and the whole arm because of the continuous and almost tonic strain on the muscles and tendons, which in the course of time results in failure of power in the right hand.

Bernardino Ramazzini, 1713,
as cited in Herington and Morse, 1995

Musculoskeletal Disorders (MSDs), a term used to describe the various disorders that may arise over time or by overexertion, are injuries and disorders of the muscles, nerves, tendons, ligaments, joints, cartilage and spinal discs. Examples include carpal tunnel syndrome, rotator cuff tendonitis, and herniated or ruptured lumbar discs. MSDs can be related, both directly and indirectly, to aspects of the work or the work environment known as risk factors. Non-work activities and environments that expose people to these risk factors can also cause or contribute to MSDs. When an MSD is associated with work it is usually referred to as a Work Related Musculoskeletal Disorder (WRMSD or WMSD). Other terms, such as cumulative trauma disorder (CTD), repetitive stress injury (RSI) and repetitive motion injury (RMI), mean roughly the same thing as MSD.

1.4 Ergonomics and Computer Work

Beginning in the 1970s, computers were introduced into the office environment as productivity tools, and have reshaped the way we work and communicate. Today, nearly all office workers spend at least part of their day interacting with a computer, and for some, the computer and its various accessories form their entire working life.

This new tool has transformed the varied writing, typing, reading, filing, meeting and communications of yesterday. The flexible array of tasks that office workers once performed are being reshaped, replaced and refocused into one primary device, the computer. Consequently, the diverse postures and dynamic nature of a day at the office of yesteryear have for many transformed into a focused, hands on the keyboard and mouse, eyes on the screen, buttocks in the chair, long duration static seated postures.

Ergonomics encompasses both physical and mental aspects of work. This book will focus primarily on the physical ergonomics issues in the computerized office environment.

Reprinted from Ergonomics Today™. Used with permission.

What is Cognitive Ergonomics?

June 11, 2001

Ergonomics is sometimes described as "fitting the system to the human," meaning that through informed decisions; equipment, tools, environments and tasks can be selected and designed to fit unique human abilities and limitations. Typical examples in the "physical ergonomics" arena include designing a lifting job to occur at or near waist height, selecting a tool shape that reduces awkward postures, and reducing unnecessary tasks and movements to increase production or reduce errors and waste. "Cognitive ergonomics," on the other hand, focuses on the fit between human cognitive abilities and limitations and the machine, task, environment, etc. Example cognitive ergonomics applications include designing a software interface to be "easy to use," designing a sign so that the majority of people will understand and act in the intended manner, and designing an airplane cockpit or nuclear power plant control system so that the operators will not make catastrophic errors.

Cognitive ergonomics is especially important in the design of complex, high-tech or automated systems. A poorly designed cellular phone user-interface may not cause an accident, but it may well cause great frustration on the part of the consumer and result in a marketplace-driven business failure. A poor interface design on industrial automated equipment, though, may result in decreased production and quality, or even a life threatening accident.

Complex automated systems create interesting design challenges, and research and post accident analysis indicate that the human role in automated systems must be closely considered. Automation can result in increased operator monitoring and vigilance requirements, complex decision-making requirements, and other issues that can increase the likelihood of errors and accidents.

Another interesting effect in automation is that humans will sometimes over-trust or mistrust an automated system.

The Three Mile Island nuclear power plant accident is in part an example of the effect of people over-trusting a system. During that event, the control panel indicated that an important valve had operated as instructed, and the control room operators trusted the system was reporting accurately. Actually, the valve had not operated as instructed, and it became a key point in the failure that resulted in a serious mishap. (Interestingly, some will blame the operators, when in fact, under the mental load created by the evolving accident, they performed as an ergonomist would expect. The actual cause of the accident is a control system design error that provided incorrect information to the operators.)

An example of mistrusting a system occurred at a medium security women's prison in Oregon, USA, when a new surveillance system was installed. The alarm was triggered whenever it sensed motion in particular areas of the facility. During the first few weeks, the alarm was repeatedly triggered by everything from birds to leaves blowing in the wind. The guards became conditioned to the fact that it often triggered in error and began to ignore it. Using this to her advantage, a prisoner climbed over the fence knowing that the alarm would go off, but that the guards would most likely ignore it long enough for her to escape. It worked. When this same mistrust effect occurs with something as important as a fire alarm, the results can be deadly.

Physical ergonomics issues, primarily in the workplace, dominate the public view and understanding of ergonomics. Fortunately, ergonomists are busy behind the scenes working to improve all human-machine interfaces, including the cognitive aspects. Unfortunately, many companies, engineers, regulators and other decision makers fail to recognize the human factor in design, and many unnecessary errors, accidents, product failures and other business costs are the predictable result.

– Peter Budnick, Ph.D., CPE and Rachel Michael, M.Sc., AEP

1.5 References and Additional Information

Hagberg, M., Christiani, D., Courtney, T. K., Halperin, W., Leamon T. B., and Smith, T. J. (1997). Conceptual and Definitional Issues in Occupational Injury Epidemiology. American Journal of Industrial Medicine. 32, 106-115.

Herington, T. N. & Morse, L. H. (1995). Cumulative Trauma/Repetitive Motion Injuries. In T. N. Herrington and L. H. Morse (Eds.), Occupational Injuries: Evaluation, Management, and Prevention (pp. 333-345). St. Louis, MO: Mosby.

Jastrzebowski, W., An Outline of Ergonomics, or the Science of Work Based Upon the Truths Drawn From the Science of Nature, originally published in Nature and Industry (1857), reprinted by the Central Institute for Labour Protection (1997). Warsaw, Poland.

2. Risk Factors, MSDs and Computer Work

2.1 Risk Factors

Risk factors are actions, conditions or a combination of these that may reduce performance, create discomfort or lead to an MSD. Examples of risk factors include:

- Awkward posture
- High force
- High repetition
- Long durations
- Contact pressure
- Environmental factors (heat, cold, ventilation, light, vibration)

2.1.1 Awkward Postures

Posture is the position of the body. Awkward posture is a relative term that indicates a joint is deviated away from its neutral or preferred position for a given task. Generally, the more awkward a posture, the greater the risk of injury and degradation in performance.

The effect of even a minor postural deviation, such as turning the head to one side, is not typically a concern for a healthy person. But when, for instance, a worker remains in such an "awkward" posture for many hours each day, day after day, week after week or year after year, serious discomfort or injury can occur.

Posture is perhaps the most important factor in office work, and will be discussed in greater detail later in this book.

2.1.2 Force

Most computer activities can be accomplished with minor force exertions. Keyboarding requires very little force to activate the keys, although some operators do exert higher forces than needed.

This book will not discuss non-computer activities in great detail, but there are some office tasks that do in fact require the application of higher forces, including stapling, paper hole-punching and stamping. Just as sitting in one awkward posture for a short time probably will not cause a problem, low volume stapling or hole punching may not be an issue in many offices. However, in a production situation were such activities are repetitive or of long duration, these tasks can be improved through ergonomics.

2.1.3 Repetition

Repetition is a measure of the frequency at which an action is repeated. For example, a typist who can type 60 words per minute might average approximately 350 keystrokes per minute.

2.1.4 Duration

Duration is the measure of time at which a person is exposed to one or more of the other risk factors. Duration can be as low as the time it takes to strike a key, or it can refer to the length of time, for example, a worker types during a typical day, or it might refer to the length of a working day, such as 8 hours.

2.1.5 Contact Stress

Pressure is a measure of force distributed over an area. The more concentrated the force, the higher the pressure. Some body tissues are more susceptible to pressure than others. For example, the tissues on the bottom of the feet are well suited for pressures, but the tissues on the underside of the forearm are not.

2.1.6 Environmental Factors

Environmental factors like lighting, noise, vibration, heat and cold can also become risk factors. For example, glare on a computer screen is a common problem caused by reflections from direct light sources like windows or overhead lights.

2.2 Risk Factor Summary

Just like smoking cigarettes is a risk factor for developing lung cancer, repetitively reaching for a mouse or forcefully striking the keyboard are risk factors for discomfort, reduced performance or developing an MSD. And just like no one is sure how many cigarettes it takes to cause lung cancer in any one person, no one is sure exactly how many repetitions, how much force or how long a duration it takes to cause an MSD.

It is known, however, that smoking cigarettes is hazardous, and scientists are comfortable suggesting that the more a person smokes, the more likely that the person will develop negative health effects. The same can be said for exposures to ergonomics risk factors. Scientists can't tell us what specific levels will cause what specific health effects (dose-response), but they do know that the higher the exposure, especially in combination with other risk factors, the more likely an injury will be.

Reprinted from Ergonomics Today™. Used with permission.

MSD Risk Factors Cost Even Before They Injure

May 3, 2002

It's one of the first things that Ergoweb teaches at our Applied Workplace Training seminars: learn to recognize risk factors for musculoskeletal disorders (MSDs). These risk factors are actions in the workplace, workplace conditions or a combination thereof, which may cause or aggravate an MSD. Examples include forceful exertions, awkward postures, repetitive motions, vibration, duration of exposure and environmental factors such as temperature.

We know that these risk factors can result in pain, discomfort, illness and injury. We know that such disorders can result in significant direct and indirect costs to companies. We know that they can cause a lifetime of pain for those who suffer from them. We also know that injured, pained and even uncomfortable workers are not likely to perform at their best.

Unfortunately, many companies never know the full cost of neglecting occupational ergonomics until well after the fact, when injuries show up as a significant, measurable loss to the company and to the injured employees. And these costs don't include the productivity, quality, waste and process error losses that inevitably occur when poor ergonomics is accepted. These financial losses are often far greater than the health and safety costs.

Recognizing and controlling MSD risk factors is a key component to workplace ergonomics. There is plenty of healthy debate about what levels of exposure to the different risk factors workers can withstand, and researchers have yet to establish widely accepted "dose-response" relationships between known risk factors and injuries. However, at least one new study documents that MSD risk factors are costing companies money even before injuries occur.

In a study published in the May 2002 issue of *Occupational and Environmental Medicine*, Dutch researchers found that risk factors like flexion and rotation of the trunk, lifting and low job satisfaction are risk factors for sickness absence due to low back pain. In most cases, pain and discomfort are not described as injuries and are therefore not recorded in injury data logs.

In another study published in the May 2002 *American Journal of Industrial Medicine*, researchers found that neck and shoulder pain and stiffness, and medical disorders linked to these symptoms, are more widespread among computer users than previous studies suggest. Again, neck and shoulder pain usually isn't recorded as an injury, but you can be sure that an employee in pain is not performing at his or her optimum.

If your organization bases its need for ergonomics solely on injury data, it may miss opportunities to improve work and process design that will head off costly injuries and illnesses before they occur, but also improve process and worker performance, reduce costly absenteeism, turnover and other wasteful inefficiencies.

Identify, prioritize and control MSD risk factors in your workplace. It's well worth the effort.

– Peter Budnick, Ph.D., CPE and Rachel Michael, M.Sc., AEP

Reprinted from The Ergonomics Report™, Vol. 1, No. 4, June 2002. Used with permission.

Do MSD Risk Factors Cost Even Without Injuries?

Researcher and Experience Say Yes

We know that musculoskeletal disorder (MSD) risk factors, like forceful exertions, awkward postures and repetition can result in pain, discomfort, illness and injury. We also know that related injuries can result in significant direct and indirect costs to companies. And we know that such injuries can cause a lifetime of pain for those afflicted. But what about exposure to levels of MSD risk factors that do not present a high likelihood of an actual recordable, costly injury?

From a productivity and quality standpoint, the same risk factors that may result in injury are very likely to interfere with a worker's performance. When there are barriers to worker performance, there will be losses in time (productivity) and in accuracy (quality, errors, waste). For example, when a part is positioned so that it requires a worker to assume an awkward posture, that posture and the motions necessary to achieve it take time – or more precisely, waste time. This is not a question of science, this is simple, measurable fact.

So, yes, the very same risk factors that can contribute to injury will definitely cost your company in other business critical metrics. However, new research suggests that non-injury related costs might also be higher than anticipated.

In a study published in the May 2002 issue of *Occupational and Environmental Medicine*, Dutch researchers found that risk factors like flexion and rotation of the trunk, lifting and low job satisfaction are risk factors for sickness absence due to low back pain. In most cases, pain and discomfort are not described as injuries and are therefore not recorded in injury data logs.

In another study published in the May 2002 *American Journal of Industrial Medicine*, researchers found that neck and shoulder pain and stiffness, and medical disorders linked to these symptoms, are more widespread among computer users than previous studies suggest. Neck and shoulder pain usually isn't recorded as an injury, but you can be sure that an employee in pain is not performing at his or her optimum.

If your organization bases its need for ergonomics solely on injury data, it may miss opportunities to improve work and process design that will head off costly injuries and illnesses before they occur, but also improve process and worker performance, reduce costly absenteeism, turnover and other wasteful inefficiencies.

Recognizing and controlling MSD risk factors is a key component to workplace ergonomics. There is plenty of healthy debate about what levels of exposure to the different risk factors workers can withstand, and researchers have yet to establish widely accepted "dose-response" relationships between known risk factors and injuries. However, barriers to worker performance in the form of MSD risk factors are costing companies money well before injuries occur, and even without injuries.

Identify, prioritize and control MSD risk factors in your workplace. It's well worth the effort.

– Peter Budnick, Ph.D., CPE and Rachel Michael, M.Sc., AEP

2.3 MSDs and Computer Work

2.3.1 Theoretical Model of Cumulative Trauma Disorder Mechanism

Hagberg et al. (1997) proposed a process model to explain the impact of risk factors on body tissues. An event such as a lifting task may produce a non-injury effect across low back soft tissues (exposure is too low for injury) and recovery occurs. Repeat event exposures may interfere with the normal recovery process to produce disproportional responses and eventually an injury (refer to the following illustration).

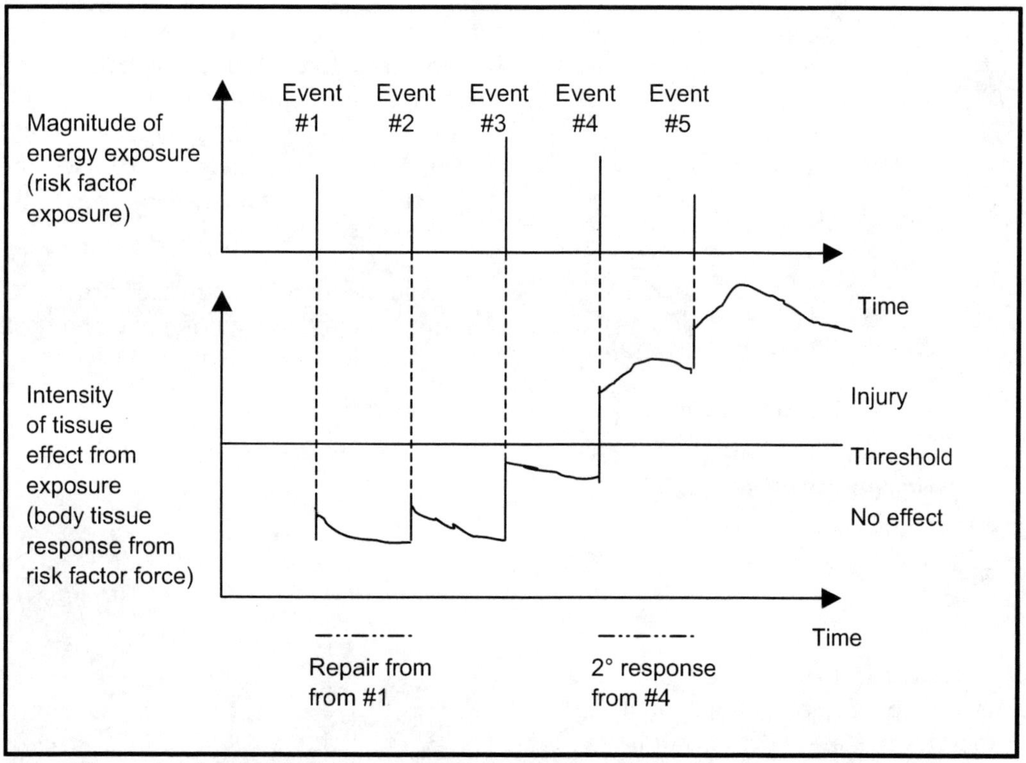

CTD mechanism model depicting impact of repeat risk factor forces on soft issues (after Hagberg, 1997)

2.3.2 Skeletal Muscle

Skeletal muscle only has the ability to contract. It is connected to bone by tendon. Muscle is composed of long, slender cells that are called muscle fibers. The fibers are bound together by connective tissue, called fascia, to eventually form the larger shape of the muscle. Nerve and blood supply are laced into the muscle. The biomechanical function of muscle is:

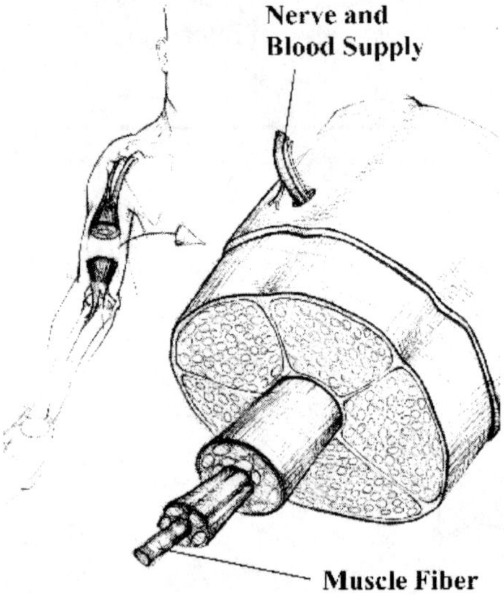

Cross-sectional view of muscle tissue showing muscle fibers

- To generate force through concentric contraction which produces body movement (i.e., shoulder muscles contract to lift the hand over the head).
- To generate force through isometric contraction which stabilizes a body part subject to a force (i.e., to maintain a standing posture, muscles in the back and neck contract to oppose gravity which tends to make the head and trunk bend forward).
- To generate force through eccentric contraction to guide motion of a body part (i.e., activity in back muscle when lowering a load).

2.3.2.1 Muscle Injuries

Two common types of occupational injury to skeletal muscle are tension neck syndrome and muscle strain/sprain. Muscle sprain/strain is unlikely in the office environment, and therefore will not be discussed here.

Tension Neck Syndrome

Tension neck syndrome is generalized muscle pain and tightness with associated small areas of taut muscle bands that are extremely tender to palpation. These small muscle spasms are sometimes referred to as a myofascial trigger point.

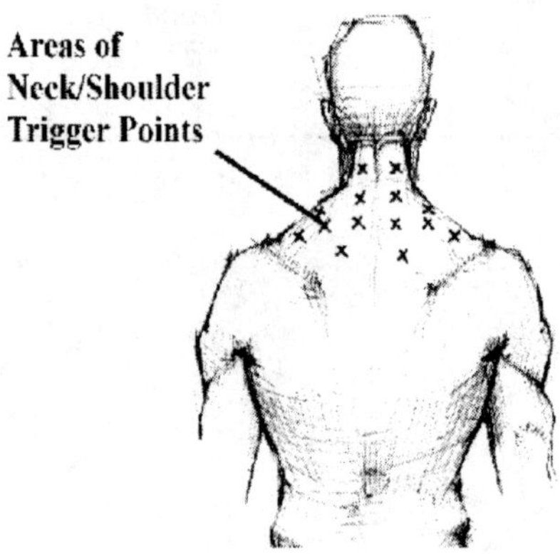

Areas of muscle tightness and trigger points

The disorder can originate from holding an awkward posture for an extended duration. Local muscle spasm and pain occurs which promotes more muscle spasm. Blood flow into the tissue becomes restricted due to the tight muscle. The resultant decrease in oxygen and nutrients coupled with the accumulation of waste products leads to pain and more muscle tightness/spasm (Herington and Morse, 1995).

Hagberg et al. (1995) reviewed the epidemiological literature and concluded that a significant association exists between tension neck syndrome and:

- Workers who perform long term static contraction such as long term looking down with the head;
- And/or workers who generate force with repetition such as repetitively looking upward or to the side.

Although the neck/shoulder and low back are common areas of muscle disorders, it should be realized that injury could occur to nearly any muscle tissue in the body depending on task demands and body motion.

2.3.3 Tendon

Tendon connects muscle tissue to bone. The force generated by muscle is transmitted through tendon to move/stabilize bone.

Tendon is composed primarily of collagen fibers that are woven together in a parallel design. Tendon tissue is enclosed by a lubricated sheath, which makes it easier for tendon to slide.

The biomechanical function of tendon is to transmit force from muscle to bone.

2.3.3.1 Tendon Injuries

Tendonitis and Tenosynovitis

The most common types of occupational injury to this tissue are tendonitis (inflammation of the tendon) and tenosynovitis (inflammation of the tendon sheath).

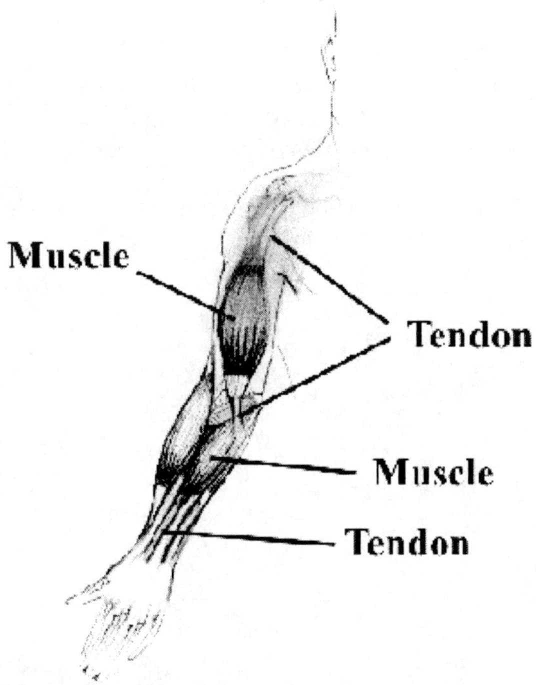

Upper extremity muscle tendon

Microtears of tendon from tension force occur daily from use of the body. Normally, they repair and no injury occurs. However, when there is repeat loading of the tendon to the extent that the microtears cannot repair, a low-grade state of pain/inflammation persists.

It is felt that tenosynovitis occurs from a friction irritation between the tendon and the sheath.

Hagberg et al. (1995) concluded that the literature shows a relationship between tendonitis and work at:

- The shoulder with repetitive and overhead activity; and
- The hand-wrist with repetitive and forceful gripping.

Occupational tendonitis has also been related to the elbow (lateral epicondylitis). As mentioned with muscle injuries, tendonitis can occur to nearly any tendon depending on task demands and body motion.

2.3.4 Spinal Disc

The spine is composed of 24 bones stacked on top of the pelvis. Between these bones are spinal discs.

The disc is made of two parts: the nucleus pulposus and the annulus fibrosus. The nucleus pulposus forms the center of the disc and is composed of water, collagen and proteoglycans. The annulus fibrosus is a series of collagen fiber layered at angles.

The primary biomechanical functions of the discs are (Norkin and Levangie, 1983):

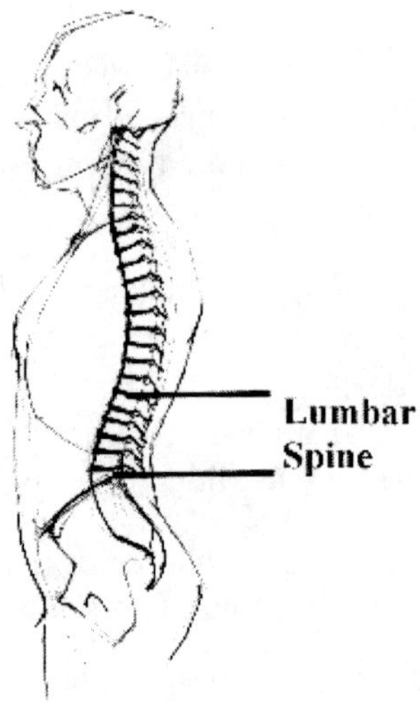

The human spine

- To resist compression and shear loading on the spine; and
- To facilitate motion of the spine by being a pliable tissue that connects moving vertebra.

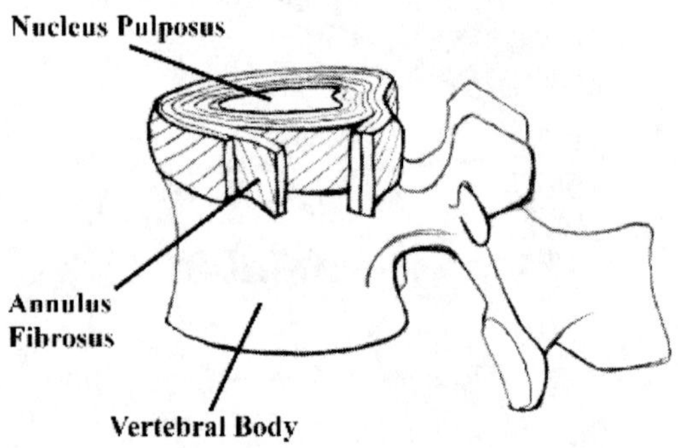

Composition and design of a lumbar vertebra and disc

The compression force going through a disc varies with body position and activity. The table below shows the compression force within the third lumbar disc as measured by Nachemson (1981) for common body positions.

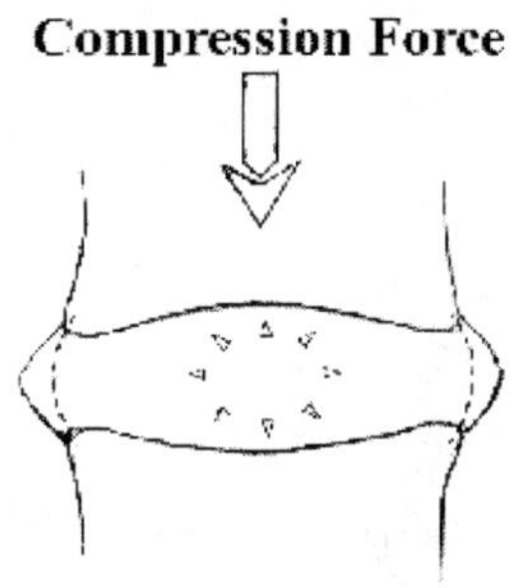

Compressive force

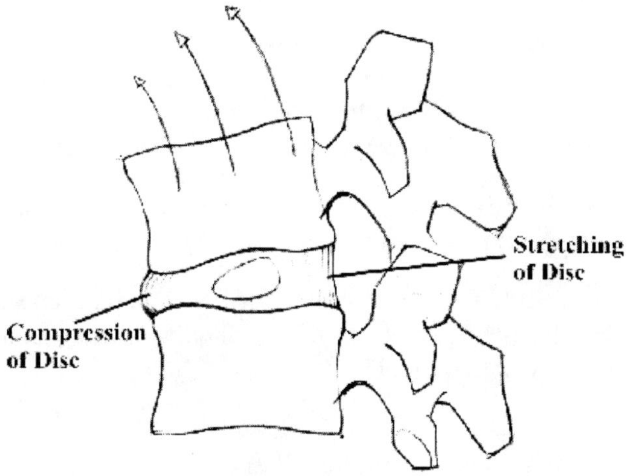

Impact of motion on spinal disc

Body Position	Spinal Compression Force (lb.)
Lying supine, awake	56
Standing at ease	112
Coughing	135
Upright sitting without support	157
20 degree forward bend	135
40 degree forward bend	225
20 degree forward bend with 44 pounds	270
20 degree forward bend and 20 degree rotation with 22 pounds	472
Lifting 22 pounds with straight back and knees bent	382
Lifting 22 pounds with back bent	427
Holding 11 pounds with arms extended	427

Spinal Compression forces measured during certain tasks

2.3.4.1 Occupational Disc Injury/Illness

Although Bigos et al. (1986) found that the dominant occupational injury/illness to the low back was strain/sprain (a finding echoed by Andersson, Fine and Silverstein, 1995), lumbar disc disorders are also significant.

Theory of Injury Mechanism

Researchers note that compression force through cadaver lumbar vertebra produces microfractures along the top and bottom borders of the vertebral body (end plate fractures) long before the disc shows sign of failure (Hutton and Adams, 1982).

Compressive forces in the range that generate these microfractures occur in the low back when certain styles of tasks (such as lifting) are performed (Chaffin and Andersson, 1991).

According to Garg (1991), "It is believed that repeated microfractures and bone scarring of the cartilage end plate leads to a weakening of the annulus fibrosus of the disc, which thus protrudes into the spinal canal."

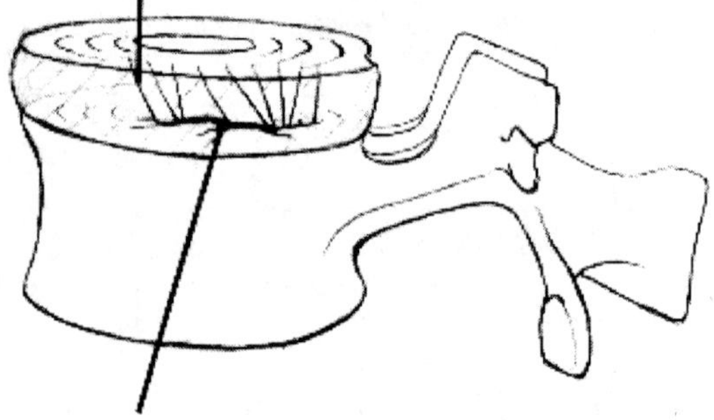

Theorized microfractures along the vertebral end plate may lead to unusual angulation of annular fibers

It is thought that annular fibers that attach to the end plate adjacent to the microfracture zones become improperly aligned. As the vertebra move, these fibers become taut and relaxed in an abnormal way, which leads to their gradual fraying and tear. This leads to increased strain on adjacent fibers and their eventual fraying and tear. Over time, this process promotes disc bulging and other local soft tissue injury.

Pope, Andersson and Chaffin (1991) used mathematical models to show that rotation of the low back produces a concentration of stress along the posterolateral aspect of the disc, a common site of disc herniation.

A strong association was shown between exposure to whole body vibration and disability due to degeneration of the intervertebral disc among crane operators (Bongers, Boshuizen, Hulshof and Koemeester, 1988). Driving motor vehicles was associated with an increased risk for acute herniated lumbar intervertebral disc (Kelsey and Hardy, 1975).

2.3.5 Peripheral Nerve

A nerve is composed of a cell body, a projection called a dendrite and a projection called an axon. A sensory nerve receives stimulus from some part of the body through its dendrite and transmits the information to its cell body and then out its axon to the spinal cord/brain. A motor nerve is stimulated by the brain/spinal cord through its dendrite and transmits the information to its cell body and then out its axon primarily to muscle. Nerve dendrites and axons can be very long (a meter or more).

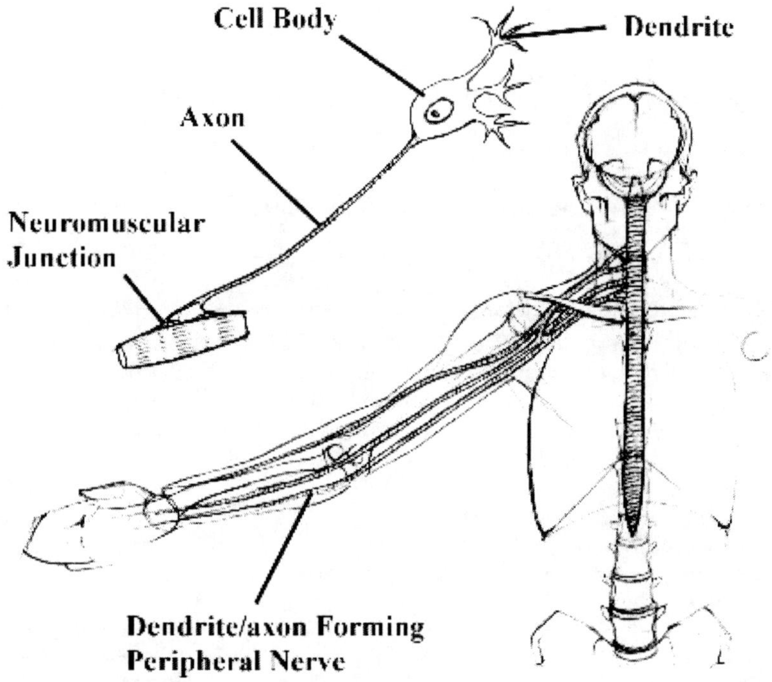

A schematic of a somatic motor nerve lies above the illustration of the major peripheral nerves of the upper extremity

A peripheral nerve is composed of the long dendrite/axon extensions of sensory and motor nerves. These fibers are wrapped in layers of connective tissue.

2.3.5.1 Occupational Nerve Disorders

Peripheral nerves can produce symptoms when under pressure or when they become entrapped (the space through which the nerve travels is too small). Refer to Section 2.4 on carpal tunnel syndrome for additional information.

Reprinted from Ergonomics Today™. Used with permission.

Ergoweb's Answer to 'What is an Ergonomics Injury?'

July 25, 2001

U.S. Secretary of Labor Elaine Chao has asked the public for input on three questions to assist OSHA in determining a direction as they struggle to deal with the issue of occupational musculoskeletal disorders. This article forms the basis of Ergoweb Inc.'s official response, and will be submitted to OSHA for the record.

Secretary Chao's first question is:

> *What is an ergonomics injury? The Department of Labor is interested in establishing an accepted definition that the Occupational Safety and Health Administration, employers and their employees can understand and apply.*

This question demonstrates a fundamental misconception about ergonomics.

There is no such thing as an "ergonomics injury."

Ergonomics is an applied, design-oriented science that, among many other things, can assist companies in reducing the occurrence of what OSHA has termed "musculoskeletal disorders" (MSDs). Applying ergonomics principles is a method through which MSDs, and their direct and indirect costs, can be controlled. Ergonomics is also applied to increase productivity and efficiency, reduce errors, improve quality, reduce waste, increase employee retention and satisfaction, and ultimately improve work, products and a company's bottom-line.

So, rephrasing the question for accuracy, Ergoweb respectfully suggests "What is an ergonomics injury?" be rephrased as "What is an MSD?"

MSDs are injuries and disorders of the muscles, nerves, tendons, ligaments, joints, cartilage and spinal discs. MSDs can be directly and indirectly related to aspects of the work or the work environment known as risk factors. Non-work activities and environments that expose people to these risk factors can also cause or contribute to MSDs. When an MSD is associated with work it is usually referred to as a Work Related Musculoskeletal Disorder (WRMSD or WMSD). Other terms, such as cumulative trauma disorder (CTD), repetitive stress injury (RSI) and repetitive motion injury (RMI), mean roughly the same thing as MSD. However, RSI and RMI are arguably inaccurate, because these terms imply that repetition is the primary risk factor, which may or may not be the case.

MSD risk factors can be defined as actions in (or out of) the workplace, workplace (or non-workplace) conditions, or a combination thereof that may cause or aggravate an MSD. Examples include forceful exertions, awkward postures, repetitive exertions and exposure to environmental factors like extreme heat, cold, humidity or vibration. It is often a combination of these risk factors that, over time, can lead to pain, injury and disability. These risk factors can be reduced through informed purchasing and workplace design, retrofit engineering controls, administrative controls, work practice definitions, or in some cases, personal protective equipment.

The manner in which a risk factor leads to an injury/disorder is usually through the accumulation of exposure to the risk factors. An event such as lifting, pushing or pulling may stress body tissues, yet the exposure may be too low for traumatic injury, and the tissues recover. Repeated exposure to this stress, on the other hand, may interfere with the normal recovery process and produce disproportionate responses and eventually an MSD type injury.

Traumatic injuries, such as an overexertion injury, may occur due to cumulative effects that manifest themselves suddenly at the time of a specific event, or they may occur because the event exposes the body to risk factors that exceed the person's individual capabilities. A sudden back or shoulder injury tied to a specific task is an example.

Ergoweb asks that OSHA refrain from using the word ergonomics in the title of any standard or guideline dealing with occupational MSDs. A more appropriate title might be "Occupational Musculoskeletal Disorder Prevention Standard" or "Occupational Musculoskeletal Disorder Reduction Standard."

Qualified ergonomists, applying ergonomics principles to the design of work, equipment and processes, is a method through which risk factors that contribute to MSDs can be effectively eliminated or reduced.

Ergonomics is not the problem; it is the solution.

- Peter Budnick, Ph.D., CPE

Reprinted from Ergonomics Today™. Used with permission.

Office Design Has Greater Impact On Women

February 10, 2003

A recent study by the Japanese International Cooperation Agency regarding office ergonomics suggests that female office workers are more likely to suffer from health problems due to poor office and workstation design than are their male counterparts.

According to an article published in *The Star* (Malaysia), 77 percent of women office workers suffered eye strain associated with computer terminals while only 63 percent of men reported similar effects. Additionally, Malaysia's National Institute of Occupational Safety and Health (NIOSH) general manager, Dr. Mat Rebi Abdul Rani, stated in the article that 69 percent of women office workers "suffered excessive fatigue from using computers too long as compared to 59 percent of male workers," and that more women than men reported wrist and upper hand pain from using computer input devices including keyboards and mice.

The study was conducted on 200 workers during a one-year period. The results, said Dr. Rani, showed that lack of ergonomic design in offices had a greater effect on women than on men. The full study will be published later this year.

Source: The Star, Malaysia

- Jeanie Croasmun

Reprinted from Ergonomics Today™. Used with permission.

NIOSH Expands Focus on Women and MSDs

November 19, 2001

The USA-based National Institute for Occupational Safety and Health (NIOSH) has an expanding research program to address the occupational safety and health needs of working women. NIOSH has identified elements of working that are of particular concern to women.

According to NIOSH:

Women currently comprise 46 percent of the 137 million workers in the United States, with their share of the labor force projected to reach 48 percent by 2008.

In 1999, 75 percent (46 million) of employed women worked full-time, while 25 percent (16 million) worked part-time.

In 1999, 3.7 million women held multiple jobs.

Sixty percent of women age 16 and over were either employed or looking for work in 1999.

Of employed women, 40 percent held technical, sales and administrative support positions; 32 percent worked in managerial and professional specialties; and 17 percent worked in service occupations in 1999.

The incidence of work related musculoskeletal disorders (MSDs) among women is one topic which will receive more research. Other factors and areas known to contribute to MSDs such as job stress, or performing patient handling in health care facilities, will also be studied.

According to NIOSH, sprains and strains, carpal tunnel syndrome (CTS), tendonitis and other MSDs account for more than half (52 percent) of the injuries and illnesses suffered by female workers, as compared to 45 percent for male workers.

Further research is needed to determine the factors that place women at greater risk for MSDs. Research will examine if physical differences between men and women, or differences in the jobs they hold, contribute to this increased risk for women. NIOSH is conducting research on musculoskeletal disorders among women in the telecommunication, health care, service and data entry industries.

In one study relating to MSDs, NIOSH worked with the Internal Revenue Service (IRS) to examine interventions for reducing discomfort among IRS data entry workers, the majority of whom are female. NIOSH found that periodic rest breaks throughout the work shift reduced musculoskeletal discomfort while allowing workers to maintain job performance.

Stress at work is another issue of concern. According to NIOSH, stress at work is a growing problem for all workers, including women. In one survey, 60 percent of employed women cited stress as their number one problem at work. Furthermore, levels of stress-related illness are nearly twice as high for women as for men.

Many job conditions contribute to stress among women, according to NIOSH. Such job conditions include heavy workload demands; little control over work; role ambiguity and conflict; job insecurity; poor relationships with coworkers and supervisors; and work that is narrow, repetitive and monotonous.

Some of these factors, such as monotonous work and input into work design, can be studied under the umbrella of macro-ergonomics. Techniques such as job enlargement and enrichment look at how to improve work that is monotonous or leaves little room for worker input.

One workplace aspect that was found to be inadequate for many female workers is the availability or properly fitting personal protective equipment (PPE). NIOSH found that much of the PPE on the market today is not designed to fit female workers. PPE manufacturers need to consider female anthropometry and design equipment accordingly.

Other areas NIOSH plans to research with regards to women in the workplace include reproductive hazards, homicide, non-fatal assault, non-traditional employment and cancers.

- *Rachel Michael, M.Sc., AEP*

2.4 Special Section on Carpal Tunnel Syndrome and Computing

Carpal Tunnel Syndrome (CTS) is probably the most widely known MSD. However, contrary to conventional wisdom, its relationship to computer work is being questioned by researchers. The following articles convey the current understanding of CTS and computer work.

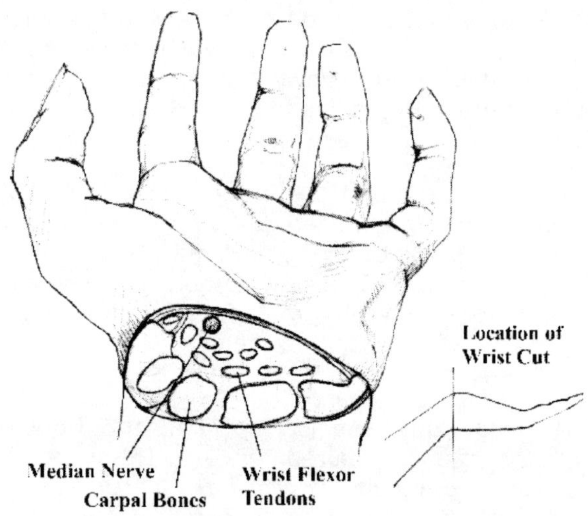

A cross-sectional view of the wrist reveals positional relationships of the median nerve, flexor tendons and carpal bones

Reprinted from Ergonomics Today™. Used with permission.

Study Questions Link Between Computer Use and Carpal Tunnel Syndrome

August 8, 2001

A study questioning the link between Carpal Tunnel Syndrome and keyboard use recently made news headlines. The authors of the study, conducted at the Mayo Clinic, Scottsdale, Arizona, reported that "the frequency of Carpal Tunnel Syndrome in computer users is similar to that in the general population." This finding surprised many, and is timely fodder for those opposed to an OSHA ergonomics regulation. To the careful observer, however, the results of this study should not be too surprising. The study provides no evidence to dispute an association between computer work and musculoskeletal disorders. It does, however, bring into question the field of medicine and healthcare professionals' ability to accurately and consistently diagnose Carpal Tunnel Syndrome.

Carpal Tunnel Syndrome (CTS) has been reported in the popular press for many years, and the medical condition has undeservedly evolved into the "poster child" for occupational ergonomics. CTS is a condition characterized by pain and numbing or tingling sensations and weakness in certain areas of the hand and is caused by compression of the medium nerve in the carpal tunnel at the wrist. Exactly what causes the median nerve compression, and whether the symptoms are always indicative of compression in the anatomical carpal tunnel, continues to be debated.

The Mayo study is interesting and has certain strengths, but it also suffers from many of the same problems that critics of the scientific basis for occupational ergonomics point to. Most significant, this study did not use a control group. The study also used a relatively small number of self-selected subjects, and did not control certain important variables, including workstation design, task or risk factor exposure, and other important factors in the relationship between a worker and a computer workstation. At best, it should be viewed as preliminary. Even though these shortcomings cast a shadow of doubt over the ability to draw reliable, statistically valid, steadfast conclusions, it is still an important contribution to the body of knowledge, and it lends evidence to what many have suspected for some time: clinically diagnosed CTS directly attributable to computer work is not nearly as common as many believed.

The reason for this appears to center on the methods used to clinically diagnose CTS. In the Mayo study, detailed and thorough methods were applied to diagnose CTS. In a typical clinical setting, where a patient may see a family practitioner or company doctor, the methods and abilities to diagnose CTS may be different, and may be subject to the "art of medicine" and the differing opinions inherent therein. The fact is, many medical practitioners diagnose CTS based on symptoms alone, and never verify their diagnosis with clinical methods like the Nerve Conduction Studies used in the Mayo study.

A careful reading of the Mayo study reveals that 27%, or 70 out of the 214 subjects, actually reported symptoms often attributed to CTS. Therefore, had diagnoses been made on symptoms alone, the results would have shown that 27% of the Mayo computer users had CTS, not the 10.5% that survived the clinical CTS verification regime.

So, what does this mean? It means that 27% of the Mayo study subjects were actually experiencing potential symptoms of musculoskeletal disorders, yet only 10.5% of those were ultimately attributed to CTS. The study makes no mention of what may have caused the symptoms that affected the other 16.5%. Further, this study does little to dispute a connection between computer use and musculoskeletal disorders, but it does bring into question the ability of the medical profession to accurately diagnose Carpal Tunnel Syndrome.

Perhaps the ensuing debate should focus on the field of health care, rather than on occupational ergonomics.

– Peter Budnick, Ph.D., CPE

Adapted from The Ergonomics Report™, Vol. 1, No. 1, March 2002. Used with permission.

Carpal Tunnel Syndrome and Keyboarding: Fact or Fantasy?

The "Background Information on CTS" article immediately following defines Carpal Tunnel Syndrome, or CTS, what non-work conditions might cause it, and how it is diagnosed and treated. But what about the potential for work activities to cause CTS? Ergoweb contacted leading experts in medicine and ergonomics to learn the most recent thinking, and here's what we discovered: CTS is definitely real, but its relationship to computer activities may not be as common as once thought. If the CTS-like symptoms reported by many computer users aren't from CTS, what are they from? As usual, "the devil is in the details." In this article, learn why CTS diagnosis is so critical and what leading experts have to say about the relationship between CTS and office work.

The root cause of CTS, which no one seems to dispute, is compression of the median nerve as it passes through the carpal tunnel in the wrist. Any number of things can place abnormal pressure on the median nerve, including increased fluid pressure in the wrist, swelling of the tendon sheaths or other tissues in the wrist, obstructions from injury or disease, and joint posture. There does appear to be disagreement and confusion, however, on how to diagnose and treat CTS, and how big a role workplace factors might play in the development or aggravation of CTS.

Many physicians rely on subjective clinical diagnosis, which they arrive at through patient reporting of symptoms and some simple examination procedures that can be performed without sophisticated diagnostic equipment. Others rely on nerve conduction studies, which they consider to be a more objective diagnostic method. Degradation in the ability of the median nerve to pass electrical current through the wrist, recorded as a slowing in nerve conduction velocity (NCV), is considered by many a positive CTS diagnosis. Even this technique is not foolproof, however. Tom Albin, an ergonomist with Alabama-based consulting firm Auburn Engineers, notes that recent research suggests that "while [NCV] is an important adjunct to clinical assessment of CTS, it has high rates of false-positive results." He cites one study that found over 18% of a control group that had no symptoms of CTS, yet tested positive in NCV tests.

Regardless of whether NCV should be considered the "gold standard" for diagnosing CTS, what some believe is commonplace are practitioners who diagnose CTS without complete examinations that eliminate other conditions that can mimic CTS. Such premature diagnosis can initiate a costly and painful chain of treatment events that ultimately not only fail, but cause additional pain and suffering, and even permanent impairment.

A recent non-peer-reviewed study by researchers at the Mayo Clinic, indicative of the debate surrounding this issue, generated an uproar when the researchers concluded that the frequency of CTS in computer users is similar to that in the general population. "We were very surprised at all the controversy," said study co-author Ben Smith, M.D., Assistant Professor of Neurology and Director of the Neurophysiology Laboratory at the Mayo Clinic in Scottsdale, Arizona. Smith suggested that their findings were not inconsistent with those in other studies, and that the results should focus attention on the diagnosis and treatment of disorders, not on ergonomics interventions. In fact, he finds great credence in the goals and outcomes of ergonomics and workplace interventions. He also states that the relationship between CTS and certain types of industrial tasks is "beyond dispute." The real take-home message is that many people are prematurely diagnosed with CTS based on incomplete evaluations, and such diagnosis can trigger inappropriate, costly and unsuccessful treatments.

Robert Szabo, M.D., agrees. Szabo, an Assistant Professor of Orthopedics and Plastic Surgery and Chief of Hand and Upper Extremity Service at the University of California, Davis, wrote the dissenting opinion in a report on workplace musculoskeletal disorders by the National Research Council and the Institute of Medicine. He argued, among other things, that the scientific understanding of CTS and its relationship to keyboarding activities in particular was inadequate to support a cause and effect relationship. According to Szabo, "Some of the worst impairment cases I've seen are the result of improper treatments based on incorrect diagnosis." He feels office work can aggravate an existing CTS condition, but is not likely to be the sole cause. He believes CTS is a multifactorial disorder, and that there is rarely one cause. Jobs that present high force and repetition, though, "like a clam shucker in Seattle," do increase the risk of CTS.

Many are critical of the Mayo study and its findings. For example, Tom Albin questions why the authors failed to discuss one of their own studies that pegged CTS rates in the general population at about 1%, and instead compared their data to a Swedish population study that indicates a 3.8% incidence rate. Citing differences in diagnostic criteria, Albin explains, "This is not an apples to apples comparison." In a separate interview, Mike Wynn, Vice President of Humantech, an Ann Arbor, Michigan, consulting firm, independently reached a similar conclusion: "This is an apple-to-bananas comparison." Sifting through the brief Mayo report, many are concluding that had an "apples-to-apples" comparison been made, their results would have shown that CTS rates were indeed higher among the computer operator study population, contrary to what the authors reported. It is hoped that these issues will be clarified in future publications.

David Rempel, M.D., Associate Professor, Departments of Bioengineering, Mechanical Engineering and Medicine, University of California, San Francisco, says "There appears to be a slightly elevated risk of CTS in keyboarding tasks, but that risk is relatively small." In order to better understand what that risk is, Rempel would like to see more studies with larger sample sizes. Further, Rempel argues that whether symptoms are due to CTS or not, a significant number of computer users are experiencing CTS-like symptoms and other disorders such as tendonitis, neck tension syndrome and trapezius muscle strain.

Dr. Jeffery Katz, Associate Professor, Department of Environmental Health, Brigham & Women's Hospital, has found a high rate of self-reported symptoms among college students, possibly signaling things to come as more people use computers as an integral part of their daily activities. Though he feels CTS is quite rare in office environments, as high as 50% of the students he studied have complaints. He argues that "these symptoms are real," and feels that duration of exposure is a key factor, with 1/7th of the students experiencing symptoms after only an hour at the computer.

Most researchers agree that CTS can be caused or aggravated by workplace risk factors, including forceful hand exertions and repetition, but are hesitant to include keyboarding tasks as presenting force and repetition levels high enough to definitively cause CTS. "There's little doubt that certain CTS cases are related to work in certain manufacturing jobs, but I think the evidence linking CTS to office work is very weak," says Stephen Moore, M.D., Texas A&M University. Moore, along with co-researcher Arun Garg, Ph.D., University of Wisconsin, has worked extensively at developing a method called the Strain Index to predict the risk of developing a hand or wrist MSD for a given job or task. Their work indicates that there may be approximately a three-fold increase in the risk of CTS for manufacturing jobs that exhibit variations of high force and repetition.

Experts caution that broader MSD concerns in office environments shouldn't be thrown out with the CTS debate. Even the Mayo Clinic study reported that nearly 30% of the computer users they studied reported symptoms of numbness or pain. They identified 10.5% of them as having clinically diagnosable CTS, but found that only 3.5% of the population tested positive through nerve conduction studies. The question remains: What about those reporting problems yet not testing positive through NCV for CTS?

Dr. Emil Pascarelli, Director of Ambulatory Care at St. Luke's/Roosevelt Hospital, Professor of Clinical Medicine at the College of Physicians and Surgeons of Columbia University, and Associate Professor of Clinical Public Health at New York Hospital Cornell Medical Center, believes that upper extremity musculoskeletal disorders are complex neuromuscular syndromes that do not necessarily originate lower in the arm where the symptoms occur. He argues that the source of the disorders is proximal - closer to the spine. "CTS is a minor problem for keyboard operators," he says. "The real problem is that there are no diagnosis for many ailments because of their multifactorial nature." His experience as a clinician and his research indicates many of the problems are related to muscle and nerve problems in the head, neck and shoulder area, including thoracic outlet syndrome.

The Bottom Line

Common computer tasks do not appear to significantly increase the risk of carefully diagnosed CTS. Symptoms that mimic CTS and symptoms of other types of MSDs, however, are elevated among computer operators. Symptoms experienced by computer operators are thought to be multifactorial in nature, making it difficult to diagnose a single root cause, and therefore an effective course of treatment. Many of these symptoms are not indicative of permanent damage or injury, and will subside with appropriate workplace modifications or medical treatments.

The most important message is that premature, incorrect diagnosis of CTS can initiate costly, painful and unsuccessful treatments, potentially ending in permanent injury or disability.

Manufacturing and industrial jobs that expose workers to high force or repetition do increase the risk of CTS and other MSDs.

– Peter Budnick, Ph.D., CPE

Reprinted from The Ergonomics Report™, Vol. 1, No. 1, March 2002. Used with permission.

Background Information on CTS

What is CTS?

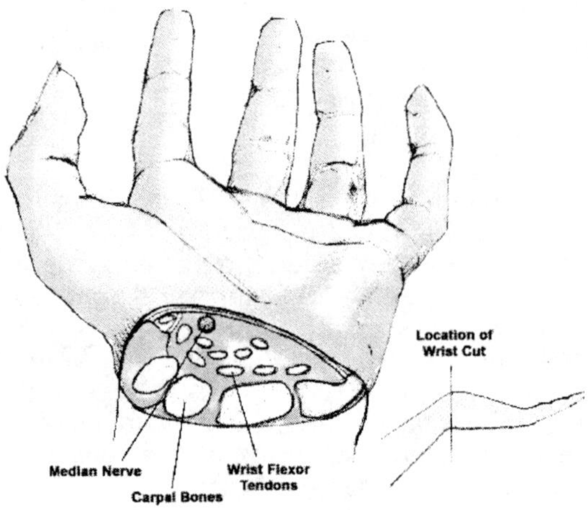

Reprinted from the *Ergonomics Image Gallery*™ with permission from Ergoweb Inc. Copyright © 2004.

"Carpus," derived from the Greek word "karpos," means "wrist." The wrist is surrounded by a band of fibrous tissue called the transverse carpal ligament, which functions as a support for the joint. The tight space between the transverse carpal ligament and the carpal bones of the wrist is called the carpal tunnel. The median nerve passes through the carpal tunnel to reach the thumb, index and middle fingers of the hand. Alongside the median nerve as it passes through the carpal tunnel are the flexor tendons, which transfer force from muscles in the forearm to the fingers. Anything that results in compression of the median nerve, such as swelling or a change in position of the tissue within the carpal tunnel, can irritate the median nerve, causing tingling and numbness of the thumb, index and the middle fingers. The nerve also supplies a branch to the muscles of the thumb, and if the condition is allowed to progress, weakness in the muscles that provide opposition force to the thumb may hinder one's grip.

What Causes CTS?

For many, the cause of their carpal tunnel syndrome is unknown. Any condition that exerts pressure on the median nerve can cause carpal tunnel syndrome. Common medical conditions that may lead to carpal tunnel syndrome include obesity, pregnancy, hypothyroidism, arthritis, diabetes and trauma. Rare diseases, including amyloidosis, sarcoidosis, multiple myeloma and leukemia, can result in deposits in and around the carpal tunnel, leading to nerve irritation.

How is CTS Diagnosed?

A doctor should perform a thorough physical exam that focuses on specific symptoms and their location in the hand, making sure they are restricted to the distribution of the median nerve. Other conditions can mimic carpal tunnel syndrome, so it's important to exclude them through physical examinations of the neck, shoulder, elbow, pulses and reflexes. Electromyography (EMG) is sometimes used to test other muscles in the shoulders, neck and arm during this process. The wrist should be examined for swelling, warmth, tenderness, deformity and discoloration. Sometimes tapping the front of the wrist can reproduce tingling of the hand and is referred to as Tinel's sign of carpal tunnel syndrome. Symptoms can also at times be reproduced using Phalen's maneuver, which involves bending the wrist forward. A CTS diagnosis is considered strong when a nerve conduction velocity (NCV) test is abnormal. The NVC test measures the speed of electrical impulses as they travel down a nerve. In carpal tunnel syndrome, the impulse slows as it crosses through the carpal tunnel. Blood tests are sometimes performed to test for medical conditions associated with carpal tunnel syndrome.

Why is Proper Diagnosis So Important?

An incorrect diagnosis can initiate a potentially serious chain of errors. Incorrect diagnosis, even by the patient, delay appropriate care, lead to inappropriate use of medications, and even encourage a doctor to order an unnecessary and potentially painful or harmful procedure or test. It is believed that many carpal tunnel surgeries are unnecessary and may result in further complications and permanent problems.

How is CTS Treated?

Treatments depend on symptom severity and the underlying root cause(s). Initial treatments may include rest, immobilization of the wrist in a splint, and occasionally ice application. Where occupational activities are suspected as a cause or aggravator, workplace modifications that reduce risk factor exposure should be pursued. Often, such modifications, sometimes in conjunction with therapy, can prevent or reduce symptoms. Depending on root causes and symptom severity, medication may be prescribed. Severe cases may require surgery. If you suspect CTS, consult with a qualified physician(s) that perform a thorough examination.

Sources:

http://www.medicinenet.com

http://www.medicalmultimediagroup.com

- Peter Budnick, Ph.D., CPE

Reprinted from Ergonomics Today™. Used with permission.

Carpal Tunnel Not the Keyboard's Fault

June 11, 2003

Researchers in Denmark have once again raised doubts that keyboard usage has an impact on developing Carpal Tunnel Syndrome (CTS), stating that, in their opinion, keyboards are not an "occupational risk for developing carpal tunnel syndrome."

Their findings are the result of a one year follow-up study of nearly 6000 Danish workers who had been surveyed previously regarding the symptom distribution and frequency of CTS.

According to a report in the *Journal of American Medicine* (JAMA), the researchers found "a prevalence of possible carpal tunnel syndrome between 1.4 percent and 4.8 percent based on a screening questionnaire and a clinical interview, and an incidence of new or aggravated symptoms of possible carpal tunnel syndrome of 5.5 percent." This, concluded the researchers, "does not pose a severe occupational hazard for developing (the) symptoms" of CTS.

The goal of the study was to determine whether or not work-related keyboard and mouse use could contribute to the development of CTS. While the researchers indicated that keyboard usage probably is not linked to CTS, they did find an association between using a mouse for more than 20 hours each week and a slightly elevated risk of developing CTS. They also noted that evidence existed that linked "forceful industrial work" to the development of CTS.

These findings will come as no surprise to readers of *The Ergonomics Report™*. The March 2002 issue reported in detail that the link between keyboarding and CTS was tenuous at best. The article concluded with:

> Common computer tasks do not appear to significantly increase the risk of carefully diagnosed CTS. Symptoms that mimic CTS and symptoms of other types of MSDs, however, are elevated among computer operators. Symptoms experienced by computer operators are thought to be multifactorial in nature, making it difficult to diagnose a single root cause, and therefore an effective course of treatment. Many of these symptoms are not indicative of permanent damage or injury, and will subside with appropriate workplace modifications or medical treatments.
>
> The most important message is that premature, incorrect diagnosis of CTS can initiate costly, painful and unsuccessful treatments, potentially ending in permanent injury or disability.
>
> – *Dr. Peter Budnick,* The Ergonomics Report™, *Vol. 1, No. 1, March 2002*

Sources: Reuters.com; The Ergonomics Report

– Jeanie Croasmun

2.5 References and Additional Information

Andersson, G. B. J., Fine, L. J., & Silverstein, B. A. (1995). Musculoskeletal Disorders. In B.S. Levy & D. H. Wegman (Eds.), Occupational Health: Recognizing and Preventing Work-Related Disease (3rd ed.) (pp. 455-489). Boston: Little, Brown and Company.

Bigos, S. J., Spengler, D. M., Martain, N. A., Zeh, J., Fisher, L., Nachemson, A., & Wang, M. H. (1986). Back Injuries in Industry: A Retrospective Study II. Injury Factors. Spine, 11, 3, 246-251.

Bongers, P. M., Boshuizen, H. C., Hulshof, C. T. J., & Koemeester, A. P. (1988). Back disorders in crane operators exposed to whole-body vibration. International Archives of Occupational and Environmental Health, 60, 129-137.

Chaffin, D. B. & Andersson, G. B. J. (1991). Occupation Biomechanics. New York: John Wiley & Sons, Inc.

Garg, A. (1991). Basis for Guide: Biomechanical Approach. Scientific Support Documentation for the Revised 1991 NIOSH Lifting Equation: Technical Contract Reports, (Chapter 3). U.S. Department of Commerce. PB91-226274.

Hagberg, M., Silverstein, B., Wells, R., Smith, M. J., Hendrick, H. W., Carayon, P., & Perusse, M. (1995). Work Related Musculoskeletal Disorders (WMSDs): A Reference Book for Prevention. Bristol, PA: Taylor & Francis Ltd.

Hagberg, M., Christiani, D., Courtney, T. K., Halperin, W., Leamon T. B., and Smith, T. J. (1997). Conceptual and Definitional Issues in Occupational Injury Epidemiology. American Journal of Industrial Medicine. 32, 106-115.

Herington, T. N. & Morse, L. H. (1995). Cumulative Trauma/Repetitive Motion Injuries. In T. N. Herrington and L. H. Morse (Eds.), Occupational Injuries: Evaluation, Management, and Prevention (pp. 333-345). St. Louis, MO: Mosby.

Hutton, W. C. & Adams, M. A. (1982). Can the Lumbar Spine be Crushed in Heavy Lifting. Spine, 7,6, 586-590.

Kelsey, J. L. & Hardy, R. J. (1975). Driving Of Motor Vehicles as a Risk Factor for Acute Herniated Lumbar Intervertebral Disc. American Journal of Epidemiology, 102, 1, 63-73.

Nachemson, A. L. (1981). Disc Pressure Measurements. Spine, 6, 1, 93-97.

Norkin, C. C. & Levangie, P. K. (1983). Joint Structure and Function: A Comprehensive Analysis. Philadelphia: F.A. Davis Company.

Pope, M. H., Andersson, G. B. J., & Chaffin, D. B. (1991). The Workplace. In M. H. Pope, G. B. J. Andersson, J. W. Frymoyer & D. B. Chaffin (Eds.), Occupational Low Back Pain: Assessment, Treatment and Prevention (pp.117-131) St. Louis: Mosby Year Book.

3. A Systems Approach to Office Ergonomics

3.1 Integrating Occupational Ergonomics Into an Organization

3.1.1 Overview

Ergonomics is a multidisciplinary field, drawing knowledge from various overlapping interests and responsibilities. Engineering needs ergonomics to properly design products and systems to maximize usability, productivity and quality. Facility Managers and Purchasers need ergonomics to ensure that purchasing decisions balance employee process requirements with capital equipment costs. Health and Safety needs ergonomics to ensure hazard exposures are understood and controlled to acceptable levels. Risk Mangers need ergonomics to properly manage company exposure and control insurance and liability costs. Medical Practitioners need ergonomics to recognize the potential source of pain and injuries, and to be able to take proper steps to treat and rehabilitate workers. Human Resources needs ergonomics to properly train and match new hires or return-to-work employees with appropriate jobs, and to attract and retain talented people. Employees need ergonomics to help them work in a comfortable and productive manner. Managers need ergonomics to ensure that corporate objectives, including profits, are met. The list goes on and on. In fact, ergonomics' greatest strength is that, when properly understood and embraced, it provides a win for everyone in the organization. This overlapping of professions, interests, departments and processes is also, perhaps, its greatest hurdle, because many organizations are not well suited for such a shared budgets and responsibility approach. Many proponents, therefore, direct ergonomics into an organization from one particular viewpoint, using the resources of one particular department. Others are able to build and launch a team approach. Others never get out of the starting gate and never realize the benefits.

It may not matter exactly how ergonomics is introduced as long as it is understood and implemented as a continuous human performance improvement process. Such a process can take the form of a program, as OSHA and others have suggested, or it can be integrated seamlessly with other business objectives, policies and procedures.

There is one key ingredient, however, that makes ergonomics work in any organization: Management support. Without it, ergonomics, like any other business initiative, is doomed to fail. With it, proponents have the ability, budgets and approvals to build and implement an action plan. Once implemented, the life and death of ergonomics depends upon demonstrating value to the organization. That value might be measured in compliance or, more likely, in improving the bottom line. In the absence of ergonomics regulations or the absence of enforcement where regulations do exist, focusing ergonomics on continuous improvement among various business metrics is the surest way to success.

Ergoweb recommends a continuous process approach like the following:

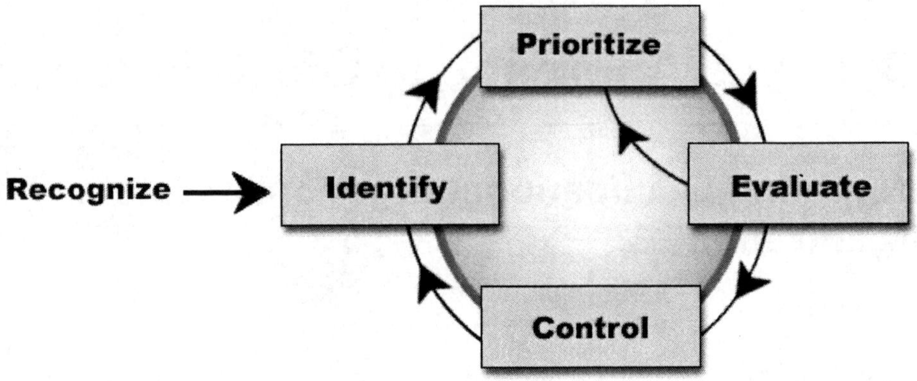

Once an organization recognizes ergonomics as an opportunity, they can begin to identify specific improvement opportunities, prioritize and evaluate those opportunities, and implement clearly understood controls and solutions. And that process repeats, because there is rarely a perfect solution, and there is never an end to opportunities for improvement.

No matter how ergonomics arrives, an organized team approach representing as many viewpoints as possible, with clear objectives, resources and responsibilities, will achieve the best results.

The federal Occupational Safety and Health Administration (1999), the National Institute for Occupational Safety and Health (1997) and the United States General Account Office (1997) have identified the following elements as basic components of an ergonomics program:

- Management leadership and employee participation
- Hazard identification and information
- Job hazard analysis and control
- Training
- Medical management
- Program evaluation

3.1.2 Management Support

As with any business goal, management must recognize and support the initiative. Presenting ergonomics in business terms is the best way to gain management support. There are three primary issues that drive all business decisions:

1. Will it increase profits?
2. Will it reduce exposure to risk or penalties?
3. Is it the "right thing to do?"

Properly understood and implemented, ergonomics is a rare example of a win-win initiative that satisfies each of these business criteria. The challenge to ergonomics proponents is to present the opportunity in clear and definable terms, quantifying the expected gains wherever possible. Good managers make informed decisions, and they rely on those they manage to provide them with the information they need to make those decisions.

3.1.3 Employee Participation

Experienced employees, those that actually perform the specific tasks and jobs, are often the best source of information about possible problems and improvements. And any continuous improvement process will benefit from an organization-wide team approach. A successful ergonomics approach should therefore include employee participation.

3.1.4 Hazard Identification and Information

From a safety perspective, people should know where potential hazards exist, and information should be available regarding how employees can minimize exposures.

3.1.5 Job Analysis and Control

Evaluating jobs and tasks to determine whether a hazard or an improvement opportunity may exist is an important component in an ergonomics process. Only when a job is understood in detail can informed recommendations to control a hazard or improve the job be made.

3.1.6 Training and Education

Ergonomics is a team approach, and each employee must understand and take responsibility in order to effectively improve. Particularly in office environments, each employee has some amount of control over the way their work areas are laid out and the methods and order in which they accomplish their daily activities. Therefore, much of the opportunity for improvement is within their control, although equipment, furniture and work tools play an equally important role.

Training should begin at the management level to ensure management recognition and support. Ultimately, all employees should receive the appropriate level of training so that they can participate in the continuous improvement process.

Training should increase awareness, explain management goals and gain employee support. Focus on increasing performance and effectiveness of all team members. Also explain that from an injury perspective, early awareness and reporting is a good thing, because it catches a problem before it develops, substantially reducing future costs.

3.1.7 Medical Management

Medical management occurs when ergonomics has failed. Ergonomics is the solution to controlling ergonomic related injuries. Medical management is the necessary reaction that supports employees that experience discomfort, pain or injury. It can range from treatment and rehabilitation to back-to-work scenarios.

3.1.8 Program Evaluation

As with all continuous improvement systems, it must be a closed loop. That is, it must continuously evaluate itself to monitor and ensure success.

3.2 Organizational Issues

Macro-ergonomics is a term used to describe a broad, systems approach to ergonomics that takes into account much more than just the physical aspects of work, people and equipment. One organizational issue that is difficult to quantify, but nevertheless important, is the psychosocial environment. This topic is important but beyond the scope of this book; however, a special section on workplace stress is presented in Section 3.6.

Larger organizations can be complex environments with many competing demands. Without a well understood and focused plan to achieve certain high level goals, an organization will likely fail, or at least fall far below its potential. The following figure conceptualizes the competing demands that must be balanced for the long term success of a company, including profits, the life-blood of any company; people, the literal life-blood of any company; and equipment and processes. Ergonomics is perhaps the most important opportunity to achieve this balance as many companies are discovering.

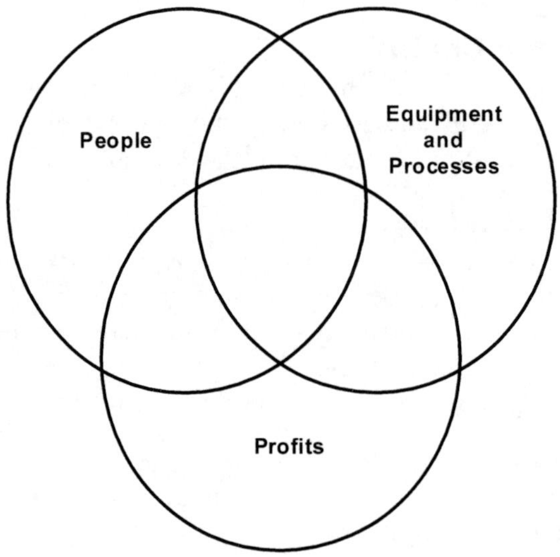

A macro-ergonomics perspective seeks a win-win balance between competing demands to maximize profits, utilize the proper equipment and processes, supply a safe, productive and fulfilling work environment for employees, and satisfy customer needs, the ultimate requirement for success.

The micro-ergonomic perspective reduces this focus to a single workstation or family of similar workstations. At this level, the balance becomes that of a worker (or group of workers), work tools and equipment like furniture, computers and accessories, and process and task demands.

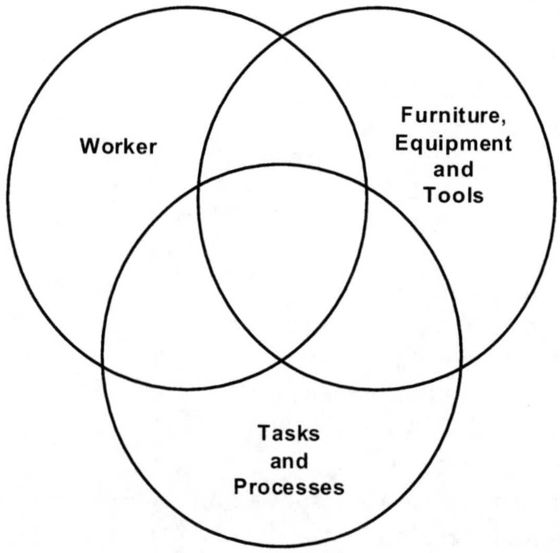

3.3 Workforce Issues

A successful company must understand and satisfy the needs and demographics of its customer base. It must likewise understand and satisfy the needs and demographics of its workforce.

People vary in all aspects. For occupational ergonomics, one of the most obvious and critical concerns is how a workforce differs in size. Anthropometry is the term ergonomists use to describe the study of variations in physical human attributes. Practical considerations regarding the differences in designing systems for individuals vs. designing for groups, or populations, will be discussed later in this book.

3.3.1 Anthropometry and Posture Evaluation

3.3.1.1 Introduction

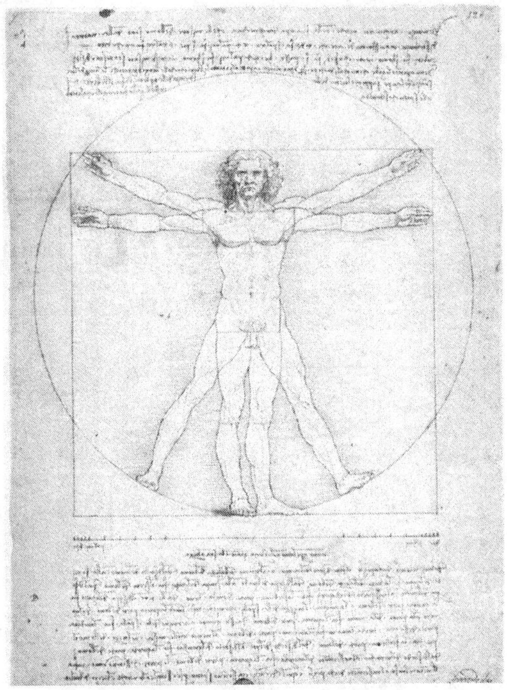

Vitrius, who lived in the first century B.C., discovered that if a person lies flat on his back, in general, the outstretched hands and feet will touch the circumference of the circle formed with the navel at the center. Leonardo Da Vinci, in his drawing 'Vitruvian Man', has immortalized Vitrius and produced one of the most recognizable images relating to anthropometry.

Anthropometry is the study of human body dimensions like height, weight, reach lengths and eye heights, and is one of the most important data sets applied by ergonomists working to fit people with their physical environment. It's easy to see that people vary in height and weight, but it's less obvious that we also vary in many other dimensions, such as leg and trunk lengths. For instance, some races have proportionally longer legs than other races, and even within the same race, specific body dimensions vary substantially from person to person.

To illustrate, imagine two women of the same height, standing side-by-side. We measure the hip height, which corresponds to leg length, and find that one has legs three inches longer than the other. Next, we measure seated knee height (popliteal height, as the dimension is technically called), and we find yet another difference between them. Finally, we ask each woman to sit in a chair and adjust the chair height so that each may comfortably rest her feet flat on the floor (shoe height introduces yet another variation). Even though they are the same standing height, we'll find that the seated height may differ substantially, because their leg and their torso lengths are different. Important dimensions like seated eye and elbow heights will also be different, which means each will have different height preferences for their keyboards, mice and monitors.

Fortunately, furniture manufacturers are finally recognizing these complex dimensional variations among people, and some are applying multivariate statistical techniques combined with user field-testing to arrive at chair and workstation designs that can comfortably accommodate large portions of the working population.

The Miss-Measure of Man

In the 19th and early 20th centuries, anthropometry was a pseudoscience used mainly to classify potential criminals by facial characteristics. For example, Cesare Lombroso's *Criminal Anthropology* (1895) claimed that murderers have prominent jaws and pickpockets have long hands and scanty beards. The work of Eugene Vidocq, which identifies criminals by facial characteristics, is still used nearly a century after its introduction in France.

The most infamous use of anthropometry was by the Nazis, whose Bureau for Enlightenment on Population Policy and Racial Welfare recommended the classification of Aryans and non-Aryans on the basis of measurements of the skull and other physical features. Craniometric certification was required by law. The Nazis set up certification institutes to further their racial policies. Not measuring up meant denial of permission to marry or work, and for many it meant the death camps.

Source: The Skeptics Dictionary by Robert Todd Carroll (www.SkepDic.com)

3.3.2 Anthropometric Data Sources

Specific anthropometric information can be derived from tables or from formulas.

3.3.2.1 Table Method

Using anthropometry tables is a simple way to determine functional measurements like reach distances. The following data is provided for reference purposes. Anthropometry will be addressed further in the Standards and Guidelines section, because furniture design standards have been developed that incorporate anthropometry concerns.

To use the tables:

- Determine which figure posture position (standing, side view seated, rear view seated) will give you the information you need.
- View the figure and observe the number of the body dimension of which you wish to know the length.
- View the table related to that figure and observe the row of values relative to that dimension on the table.
- Determine which values are significant to you based on gender and population percentile.

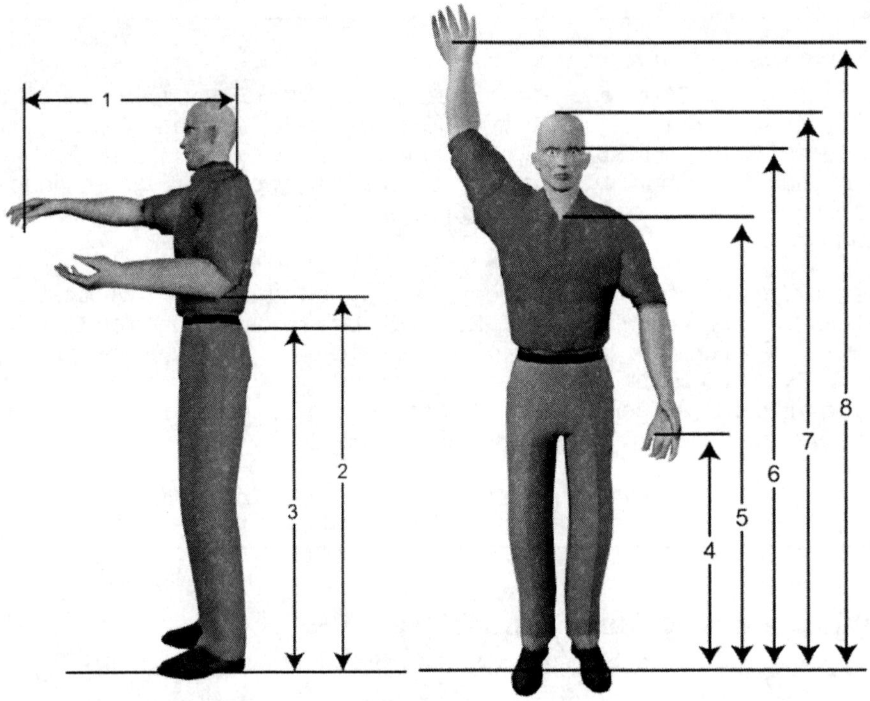

Estimates for United States Adults (in inches)

Number	Measurement	Males			Females		
		5th	50th	95th	5th	50th	95th
Side View Standing							
1	Forward functional reach	28.6	30.9	33.3	25.8	28.0	30.0
2	Elbow height	40.2	43.5	46.9	37.2	40.2	43.1
3	Waist height	38.4	41.9	45.4	36.7	40.0	43.3
Front View Standing							
4	Knuckle height	27.6	30.1	32.7	26.4	28.7	31.1
5	Shoulder height	52.4	56.7	61.0	48.2	52.2	56.1
6	Eye height	62.8	67.3	71.9	55.9	60.0	64.2
7	Stature	64.6	69.1	73.6	59.8	64.0	68.1
8	Functional overhead reach	76.8	81.9	87.0	71.1	75.8	80.5

Values for the table data were derived from Pheasant (1986) and OSHA's Working Draft of OSHA's Proposed Ergonomic Protection Standard, 1995.

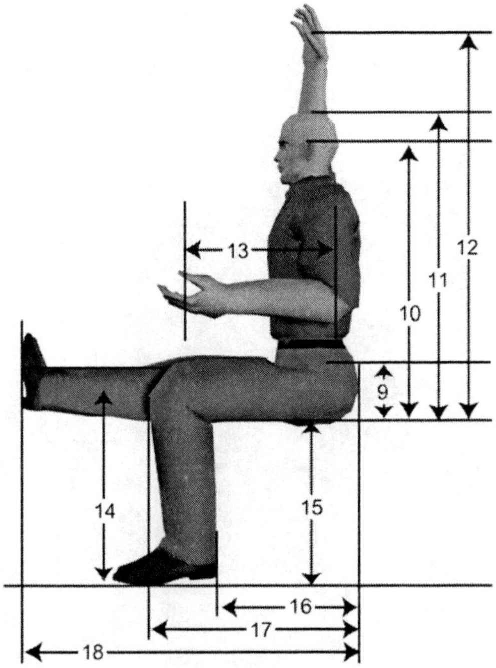

Estimates for United States Adults (in inches)

Number	Measurement	Males			Females		
		5th	50th	95th	5th	50th	95th
Side View Seated							
9	Thigh thickness	5.3	6.3	7.3	4.9	6.1	7.3
10	Eye height	29.1	31.5	33.9	27.2	29.5	31.9
11	Sitting height	33.7	36.0	38.4	31.5	33.9	36.2
12	Functional overhead reach	45.5	49.4	53.3	42.1	45.7	49.2
13	Elbow to functional reach	12.7	14.2	15.7	10.9	12.7	14.5
14	Knee height	19.5	21.7	23.8	18.1	19.9	21.7
15	Popliteal height	15.6	17.5	19.5	14.2	15.9	17.7
16	Buttock-popliteal length	17.5	19.7	21.9	17.3	19.3	21.3
17	Upper leg length	21.6	23.4	25.2	21.0	22.6	24.6
18	Leg length	38.3	41.4	44.5	36.2	39.6	42.4

Values for the table data were derived from Pheasant (1986) and OSHA's Working Draft of OSHA's Proposed Ergonomic Protection Standard, 1995.

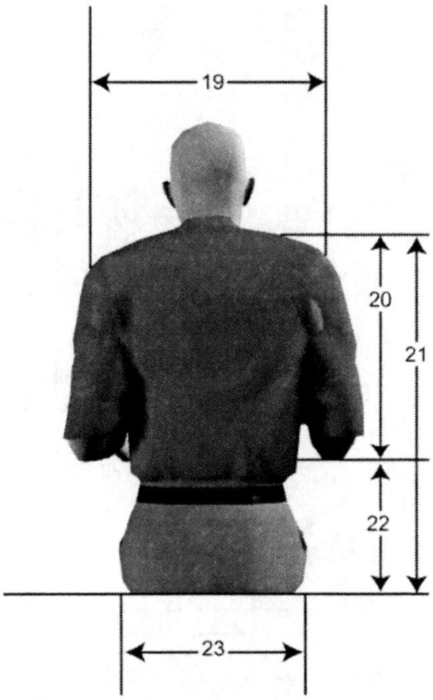

Estimates for United States Adults (in inches)

Number	Measurement	Males			Females		
		5th	50th	95th	5th	50th	95th
Rear View Seated							
19	Shoulder breadth	14.4	15.7	17.1	13.0	14.2	15.4
20	Upper arm length	13.0	14.4	15.7	12.0	13.2	14.4
21	Shoulder height	21.5	23.6	25.8	20.1	22.2	24.4
22	Elbow height	7.7	9.6	11.6	7.3	9.3	11.2
23	Hip breadth	12.2	14.2	16.1	12.2	14.8	17.3
—	Body weight (lb)	128.8	183.4	238.0	95.8	146.3	196.8

Values for the table data were derived from Pheasant (1986) and OSHA's Working Draft of OSHA's Proposed Ergonomic Protection Standard, 1995.

3.3.3 Application of Anthropometric Data

Three criteria should be determined prior to using anthropometric data:

- Define the significant body dimensions for the particular application.
- Define the task(s) that will be performed.
- Define the population that the design must fit (designing for an individual, designing for a workforce, designing for extreme sizes – 95th percentile male, 5th percentile female).

The important body dimensions are determined by the nature of the job that is being performed. A task analysis should be done to identify the job methods. This will disclose the body motions and positions required for job. This, in turn, determines what body dimensions are significant. By making use of anthropometric data it is possible to design so that relevant components are placed within their optimum areas for the given individual or population.

3.3.3.1 Defining the Population

The designer needs to consider the gender (male, female or both) and the race of the people who are going to be performing the work or using the tool.

3.3.3.2 Defining the Design Philosophy

At times only one individual needs to be considered with regard to job or equipment design. When this is the case the significant body measurements can be taken directly from the person (or from anthropometric data sources) for design specifications.

At other times a decision has to be made regarding the size of the people that should be accommodated by the tool/job. For a given gender and race, the distribution of body size (e.g., height, weight, arm length, thigh length) follows a bell curve (normal curve). The greatest incidence occurs at the 50th percentile with decreasing incidence at the extremes (5th and 95th percentiles). For United States females, the 50th percentile height is 64.0 inches, 5th percentile height is 59.8 inches and 95th percentile height is 68.1 inches (Pheasant, 1986). This means that the average height for this group is 64.0 inches. Five percent are shorter than 59.8 inches while 95 percent are taller than 59.8 inches. Ninety-five percent are shorter than 68.1 inches while 5 percent are taller than 68.1 inches. The following graph of adult female height frequency distribution illustrates this point.

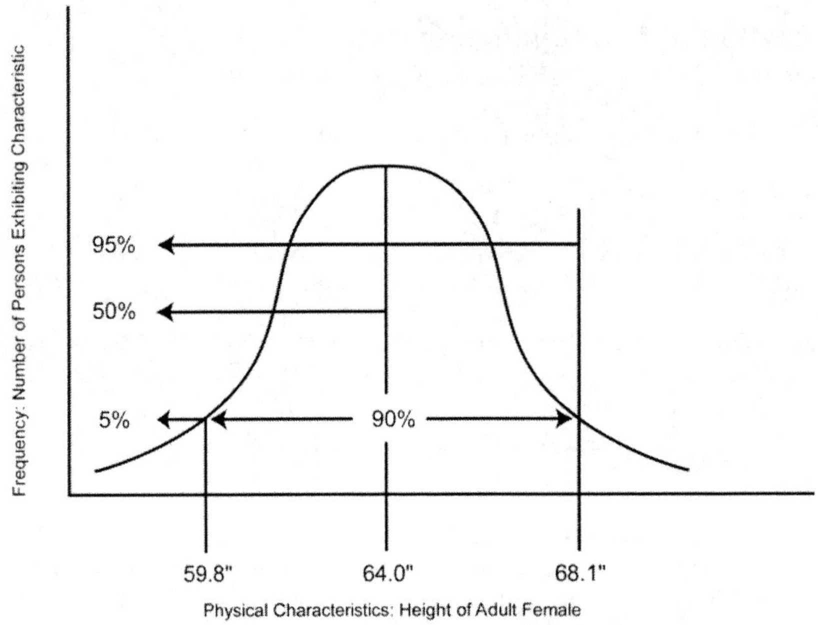

Frequency Distribution of US Adult Female Height

The designer selects the percentile that best suits the situation being considered.

When a general population needs to be accommodated, ergonomists typically design for people in the 5th to 95th percentile range, meaning people within that size or strength range will be considered in the design such that they are able to "fit" and safely and efficiently interact with the system and it's components. This often leads to adjustability features in the design, such as an adjustable chair or desk.

On occasion, adjustability is not possible. Designing for the "average" individual is criticized from the perspective that there is no "average" sized person, and designing for such may make 50 percent of the population suffer (Panero and Zelnik, 1979).

Since a design based on "average" body dimensions could cause injury or interfere with tool use/job function, "extreme" body dimensions are used. Konz recommends a general rule, "design so that the small woman can reach and the large man can fit".

The following article captures the complexity of designing for varying human populations and references some of the leading current projects and applications.

Reprinted from The Ergonomics Report™, Vol. 1, No. 5, July 2002. Used with permission.

Big, Tall, Short, Small

What's a Designer to Do?

Designers are often frustrated by humans. It sounds funny, because nearly all designs are destined to be used (or misused) by people, but it's true. Designing something to accommodate the wide variety of human sizes and shapes can be so frustrating, in fact, that many designers simply neglect to do it. Some neglect to accommodate intentionally, others try yet make mistakes in the process, and still others neglect it out of pure ignorance.

Intentionally choosing not to design a product for each and every person in the world is common, necessary and acceptable. Informed businesses, policy makers, designers and ergonomists do it and recommend it every day. Trying to accommodate a variety of people, yet failing, can be due to either misapplication of available knowledge, lack of access to available knowledge or gaps in that knowledge base. Although it occurs more often than it should, there is no excuse for failing to accommodate out of pure ignorance.

Anthropometry is the knowledge base we apply when trying to understand and accommodate the physical differences among people. Dr. Bruce Bradtmiller, President of Anthrotech, Inc., gives the simple example of how one might design a doorway using such data. If we can fit the expected height and width of people, we have the two dimensions we need to proceed. Sounds simple, right?

Actually, even such simple problems are more complicated than they appear at first glance. For example, we would want to consider allowances for clothing, shoes, hats, safety and emergency clearances, room for items that may be carried through the door, etc.

For one or two dimension design problems like the door example, we typically turn to simple "percentile" data sets. For example, if we decide to fit the height of 95 percent of all Americans through our doorway, we would select 95th percentile American height, plus allowances, as our design dimension.

What if we're designing something that has more than one or two important dimensions – an office chair, for example? We'd want to fit the length of peoples' lower legs, including shoes, to determine seat height; we'd want to fit the length of peoples' upper legs to determine seat pan depth; we'd want to fit the width of peoples' buttocks and hip width to determine seat pan width; and so on. Unfortunately, we can no longer rely on the percentile data we used to design our doorway, because we now have multiple dimensions, or multiple variables to consider, not just height and width. If we hope to achieve a good fit, we need to turn to multivariate anthropometry methods which rely on statistical approaches to account for variability among the multiple dimensions. In other words, multivariate approaches help us account for the fact that people who are, for instance, 50th percentile "average" height, do not necessarily have 50th percentile buttocks widths. If we assumed they did, we'd likely end up with a design that didn't fit anyone well. (Chair manufacturers have discovered that it is indeed difficult, or impossible, to provide a good fit for everyone with just one chair model. Many have made the informed business decision to offer big, medium and small versions to better accommodate more people.)

Researcher Kathleen Robinette of the U.S. Air Force Research Laboratory explains that multivariate anthropometry methods provide a methodology through which we select the most important dimensions for a specific product based on the intended user population, the tasks those users will be expected to perform, and environment of use. "Multivariate methods allow us to consider two or more dimensions simultaneously," adds John Roebuck, President of Roebuck Research and Consulting.

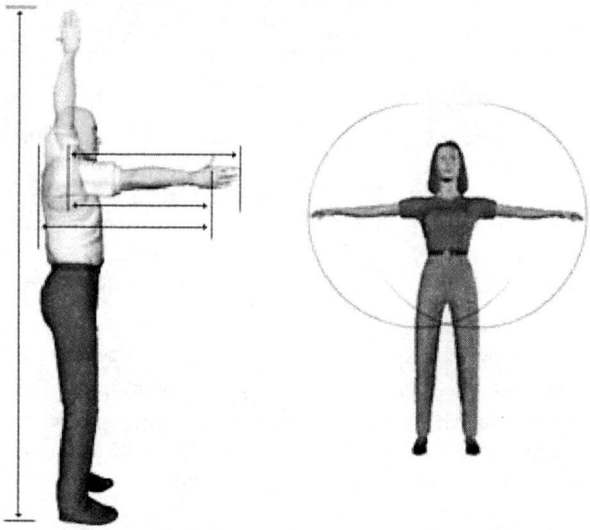

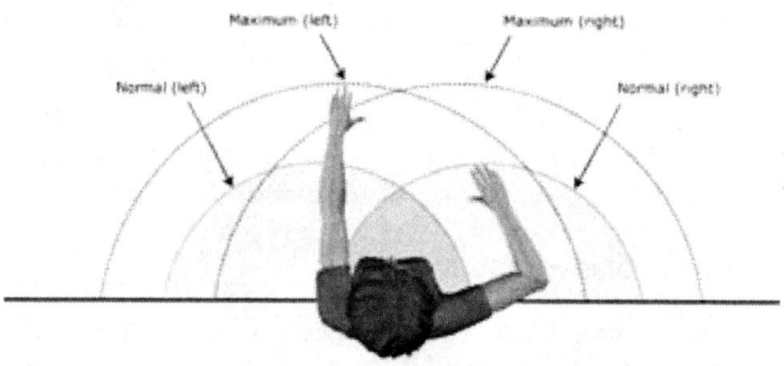

It's good to know that there are data and techniques available to help designers better fit the people they design for, but finding the data is not always easy. Sometimes, in fact, it doesn't exist. Or, the dimension thought to be appropriate turns out to be wrong.

For example, "Women's hip circumference is usually largest down on the thighs," says Robinette. "Women consider that legs, not hips, but that's where most are widest. For men, it's higher up, actually on the torso, not the thighs."

Americans Are Getting Fatter

"The historic data indicate we were getting taller until about the mid-1970s," says Robinette. "Now we don't seem to be getting taller, but we do seem to be getting fatter."

According to some estimates, about 26% of American adults are obese. That's about 54 million adults that are roughly 30 pounds or more over what is considered a healthy weight.

There's a similar trend among American kids. The percentage of significantly overweight black and Hispanic children has more than doubled in 12 years. Among white children, there was a 50% increase. As of 1998, among kids 2 to 12 years old, almost 22% of blacks, 22% of Hispanics and 12% of whites were overweight.

According to Bradtmiller, deciding who to accommodate in a design is a business or policy decision. Recent articles in the news demonstrate the growing consequences weight-based design decisions may bring. For instance, Southwest Airlines made its business decision clear when it announced a policy requiring the purchase of two tickets if the customer tends to spill over into their neighbor's seat. In fact, most economy class air travelers will report discomfort in airplane seating, either due to their own size, the size of neighboring passengers or simply due to the seat design itself. It seems that nobody feels a good fit in a typical airplane seat. If the obesity trends continue, we can speculate it won't get any better unless airlines pay more attention to – and spend more money on – seat design. And if they provide more space to passengers, they will be able to fit and sell less seats per flight, increasing the economic impact. So, especially with the airline businesses experiencing an economic downturn, don't expect the situation to get any better soon. That is, unless you are willing to pay for the increased costs, which the airlines are currently banking you won't.

Another example, reported in *Ergonomics Today*™ (November 2001), described the events surrounding a woman's death at Knott's Berry Farm, an amusement park in California, USA. According to the article, a 40-year-old woman weighing nearly 300 pounds fell to her death from the "Perilous Plunge" water ride. The manufacturer believes the woman fell from the ride because she was not properly belted and should not have gotten on the ride if she could not be safely restrained. The family's attorney claims that abrasions on her thighs indicate that she was belted in before she fell. The coroner reported that her abdomen was eight inches larger than the 50-inch seat belt, and investigators speculate that if belted, she may have placed the belt below her abdomen.

And then, a June 2002 article in *Ergonomics Today*™ summarized a study published in *Accident Analysis and Prevention* that found heavier people are more likely to be killed or seriously injured in car accidents than lighter people. Specifically, people weighing between 100 and 119 kg (220-262 lbs) are almost two-and-a-half times as likely to die in a crash as people weighing less than 60 kg (132 lbs). The study concluded that it's not just total weight, but obesity that increases the risk.

Theories abound to explain this finding. Some suggest it shows that crash tests, and therefore car interior designs, do not accurately reflect the people who may be in the car, because crash tests typically use dummies that represent a "standard-sized" male weighing about 78 kg (172 lbs). Others suggest that obese people, with health problems such as high blood pressure or diabetes, could be finding it tougher to recover from injury. Still others think the link might be due to poor seat belt restraint during a crash. One researcher suggested that a large amount of fat tissue between the restraint system and the bony thorax introduces "slack" into the restraint system, decreasing its performance. (Interestingly, among five health risk measures recently analyzed by Behavioral Risk Factor Surveillance System, obesity was up in all 47 states surveyed, and seat-belt use was up in 39 of the 47 states.)

The Bottom Line

Regardless of the reasons, it seems that designers cannot ignore the fact that at least some American populations are getting more obese. Obesity trends undoubtedly have health implications, but what implications might they have for designers? Do we change our physical world - chairs, desks, tools, etc. - to fit changing human dimensions, or do we change people to fit the physical world? In other words, do we put obese people on weight control regimes, or do we recognize these trends and get busy designing for them?

The answer, of course, depends. It really comes down to a business or policy decision. Public health officials may argue that society should do what it takes to get people to stay fit and trim. But, in civil societies there are few, if any, options available to *force* people to do something against their will. A civil society cannot force people to lose weight. A decent employer or government cannot *force* people to exercise and eat less. The only way to successfully accomplish goals that rely on human behavior is to influence individuals to take it upon themselves to, in this case, lose weight, and keep that weight off. In the mean time, designers cannot sit back and wait for people to lose weight. It is his or her job to adapt systems to humans, not vice-versa. Persuading people to be better physically fit is laudable, but it is not an acceptable approach to ergonomics on its own merit.

Design discrimination happens every day, will continue to happen, and should continue to happen. Ergonomists, engineers, architects and other designers routinely make design decisions that discriminate. If it's informed design, there's nothing inherently wrong with it, and it's usually unavoidable.

Designers and organization who neglect ergonomics principles - and unfortunately there are many - may unintentionally discriminate against a large number of users because they don't adequately understand or consider human variations in their designs. But even when ergonomics is thoroughly considered it's typical to design only for a reasonable segment of the expected user population. How to reduce risk after a design is complete, such as user and use restrictions, training, warnings, etc., is another subject directly impacted by ergonomics, but beyond the scope of this article.

If American populations are growing in girth, can your organization afford to ignore the trend? It may be time to closely evaluate business and policy decisions that are impacted by obesity.

- Peter Budnick, Ph.D., CPE

Thanks to Bruce Bradtmiller, Ph.D., President of Anthrotech, Inc. (www.anthrotech.com), Kathleen Robinette, Director of the CAESAR 3D Anthropometry Project and Principle Research Anthropologist, Air Force Research Laboratory, John Roebuck, President of Roebuck Research and Consulting, and Rachel Michael, an ergonomist with Ergoweb Inc., for assisting with this article.

3.4 Task Requirements

Another critical factor in matching people with equipment and processes is the task that each will perform. In fact, it is impossible to achieve an "ergonomic design" without thorough consideration of the specific tasks in question. For example, there are significant differences between the needs of a "creative" office worker, who has great flexibility in work habits, and a data entry or call center employee, who is restricted to a very specific set of duties. There is no "one size fits all" approach to equipment selection and process design, and each of the various tasks in an organization must be considered separately.

3.5 Furniture, Equipment and Work Tools

The following article summarizes the challenges of matching people with their physical work environment in an office.

Reprinted from Ergonomics Today™. Used with permission.

Is an "Ergonomic Chair" Enough?

July 10, 2001

A golden rule in ergonomics is "one size does not fit all," and adjustability is a key feature that can transform a typical chair into an "ergonomic chair." But, as many companies have learned the hard way, an adjustable chair is not always enough.

Successful ergonomics requires a systems approach. An office chair is one component in the office system, and the person that sits in it is another. The stapler, telephone, computer, mouse, keyboard, desk and other work tools are additional components in the system. How the person interacts with those components as he or she performs required work tasks is yet another part of the overall system. Selecting a chair without considering the rest of the system components and work tasks usually leads to a misfit and may introduce new, unforeseen problems into a workstation.

In order to fit the broad working population, a computer workstation must have multiple component adjustments, not just chair adjustments. If the chair is the only easily adjustable feature, most people will still not be able to achieve their most efficient, productive, safe and comfortable working postures. This problem is not always obvious to the casual observer, and it therefore persists in many companies.

Anthropometry, the measurement and study of human body dimensions like height, weight, reach lengths and eye heights, is one of the most important data sets applied by ergonomists working to fit people with their physical environment. It's easy to see that people vary in height and weight, but it's less obvious that we also vary in many other dimensions, such as leg and trunk lengths. For instance, some races have proportionally longer legs than other races, and even within the same race, specific body dimensions vary substantially from person to person.

To illustrate, imagine two women of the same height, standing side-by-side. We measure the hip height, which corresponds to leg length, and find that one has legs three inches longer than the other. Next, we measure seated knee height (popliteal height, as the dimension is technically called), and we find yet another difference between them. Finally, we ask each woman to sit in a chair and adjust the chair height so that each may comfortably rest her feet flat on the floor (shoe height introduces yet another variation). Even though they are the same standing height, we'll find that the seated height may differ substantially, because their leg and their torso lengths are different. Important dimensions like seated eye and elbow heights will also be different, which means each will have different height requirements for their keyboards, mice and monitors.

Contrary to claims otherwise, there is no perfect way to sit, nor is there a perfect way to arrange all workstations, even if the same job is being performed at each. Natural variability in people dictates that the system must be flexible in order to meet the unique needs and dimensions of all the different people that may interact with the system. An adjustable chair provides some of that flexibility, but not enough to accommodate the full, multifactorial spectrum of anthropometric variations. Simply put, the location of the computer keyboard and mouse need adjusting independent of the chair, as may the monitor.

The extent to which a system should be adjustable depends upon the nature and diversity of tasks to be performed at the workstation, and whether one individual person or a population of workers is being fitted to the system. If only one person will use the workstation, a one-time set-up and custom adjustment may suffice. If the workstation will be used by different people (e.g., shift work or a shared nursing station), a system that can be easily adjusted by each worker is recommended. Even without shift work, employee promotions, transfers and turnover will result in different people using the same system, and adjustment will still be needed but at a lower frequency.

There are several common and relatively inexpensive workstation components designed to accommodate variations in size and preference: footrests, keyboard and mouse platforms, and monitor arms or risers. When work surface or monitor heights are "too high," a person may benefit from a footrest, allowing her to raise her seat, yet still be able to comfortably rest her feet. When desk height, and therefore the keyboard and mouse that rest on it, is "too high" or "too low," an adjustable keyboard and mouse platform may help. And when the monitor height is "too high" or "too low," monitor adjustments may help. Each of these accommodations have strengths and weaknesses.

- **Footrests:** Relatively inexpensive and easy to use. Drawbacks include the potential for clutter under a desk, or even a tripping hazard, and may restrict freedom to move and achieve desirable dynamic postures throughout a workday. On the other hand, a footrest is a simple way to accommodate smaller people in particular, who often have difficulty keeping comfortable leg postures. If a footrest is selected, adjustable height and angle are recommended.

- **Keyboard and mouse trays:** These are available in a variety of shapes and sizes, and with a variety of adjustments. To be effective, the tray must be large enough to accommodate both the keyboard (and palm rest if desired) and the mouse, side-by-side, and on the same level. Easy height adjustment and positive and negative tilt are recommended so that the user can reduce wrist deviations in a comfortable arm posture. There are drawbacks with such platforms. For example, they usually extend out from under the desk surface, pushing the worker further away from the work surface, and they can present an obstruction under the desk, sometimes hindering leg access and movement and interfering with chair arms. Quality is important when selecting these devices; be sure that the person can type and operate the mouse with stability (i.e., avoid "flimsy" models).

- **Monitor arms and risers:** Monitor heights can be raised cheaply by using such things as old phone books. If a more tasteful solution is desired, there are a variety of monitor raising products on the market, including stackable trays or adjustable arms. Lowering a monitor is limited by the work surface height and monitor design, unless a recessed monitor desk is selected. The advent of the flat panel display makes monitor placement and adjustment easier.

Ergonomists often feel that these types of accommodations are at best a retrofit to a system with deeper problems. Ultimately, the best solution is to provide an easily adjustable work surface, not just adjustable add-ons. In fact, work surfaces that adjust through a height range from seated work to standing work are no longer uncommon. Such systems allow workers to adjust to different heights for different tasks and different people, and to make postural adjustments for comfort and fatigue over the course of the workday.

In many cases, cost is the deciding factor when selecting between add-ons or fully adjustable work surfaces. However, more manufacturers are offering fully adjustable systems or work surface adjustment retrofits, and competition is driving prices down. Furthermore, a larger upfront investment can result in significant future savings.

"Work tools" is a general term that describes the accessories that support a worker's duties. A telephone, stapler, document holder or even a writing pen are all examples of work tools in an office environment. In the same sense that a carpenter needs proper tools to perform quality woodwork in a productive manner, a call center worker needs the right tools for his or her job. Proper tool selection can be the difference between success and failure, both for the company and the individual.

- Peter Budnick, Ph.D., CPE

3.6 Special Section on Workplace Stress

Over the past few years there has been a growing international emphasis on stress in the workplace. Some countries are moving toward regulations to address workplace stress, and though such laws seem unlikely in the U.S., understanding the origins of these international efforts and their potential impact on the workplace is instructive. The following *Ergonomics Today*™ articles (Ergoweb's online ergonomics news service, available at www.ergoweb.com/news), chronicles this trend.

Reprinted from Ergonomics Today™. Used with permission.

Physical, Psychosocial and Work Organization Factors on Injury/Illness Absences

May 18, 2001

A recent study done by researchers at the Karolinska Institute in Sweden shows that work absence can be related to psychosocial issues as well as physical risk factors. The study, published in the March 2001 edition of the journal *Occupational and Environmental Medicine*, examined 3470 workers in the Swedish Postal system. Worker positions included administrators, cashiers, mail handling staff, rural postmen, office personnel, office cleaning staff, computer personnel and technicians.

The study found that certain physical, psychosocial and organizational factors were important determinants of incidence of sickness, independent of each other, and that some of the associations were gender specific.

The study quotes previous research that concludes conditions at work, such as heavy, physically arduous or monotonous tasks, increase the risk of ill health, and that the grade of employment, work organization and the psychosocial work environment may also be important for health and well being. The background research also suggests that high job demands in combination with low control can be associated with coronary heart disease and musculoskeletal disorders.

The study results found that for female workers, the strongest association with sickness absence was related to working in a forward bent position. Women reporting such problems had a more than doubled risk of being in the group with high incidence of sickness compared with women who did not report these problems. Women who performed heavy lifting or monotonous movements also showed an increased risk of sickness absence. Also, the occurrence of bullying at the workplace almost doubled the risk of being in the group with high incidence of sickness. Other factors associated with a high absence rate included seldom or never being able to discuss with the supervisor and working through illness. According to the authors, this corroborates previous findings that a forward bent position was the most important factor in low back injury symptoms in female nurses.

For males, the greatest association with sickness absence was anxiety of reorganization. The risk of being in the group with a high incidence of sickness was twice that of men who did not worry about reorganization. Men also saw heavy lifting, monotonous tasks and noise exposure increase their risk of sickness absence.

For both groups, factors reflecting low control (for example, no possibility of taking or making a private telephone call) showed an association with high incidence of sickness.

Several other reports have also pointed to psychosocial issues as a risk factor for increased injury and illness. In December 2000, Ergoweb reported that according to CCH Inc., a human resource firm, one out of every five unscheduled absences is the result of worker stress, costing business about $600 per employee every year. Studies indicate that between a quarter and a third of the work force is reporting high levels of stress, and that those reporting stress as a reason for unscheduled absenteeism has jumped 316 percent since 1995.

Two studies published in the April edition of the *Journal of Occupational Health Psychology* report that long days and pending layoffs can affect both your mental and physical health. The authors of one of the studies conclude that the employing organizations need to consider the effects that job security and job satisfaction play in the worker's overall health and safety.

The Karolinska Institute study offers further commentary on current working conditions, stating that although the Swedish work environment has changed during the past decade due to new technology and new types of work, employees still perform strenuous work tasks.

This observation is supported by a recent survey from the European Foundation for the Improvement of Working and Living Conditions which shows that while people assume that increased technology has improved the working conditions, there is no such automatic improvement or trend in any such direction.

– Rachel Michael, M.Sc., AEP

Reprinted from Ergonomics Today™. Used with permission.

Stress and Fatigue in the Workplace

September 15, 2001

We have all felt the unproductive effects of stress or fatigue on our work. For most of us it seems 'common sense' that our work quality and productivity will be compromised if we are feeling stressed that day, but what are companies doing to combat this? Do they even recognize the problem?

Safety and Health Agencies in the UK and New Zealand have compiled research, issuing statements and guidelines for combating stress and fatigue in the workplace.

The New Zealand Occupational Safety and Health Service (OSH) says that working for long periods without sleep can leave workers so fatigued that it has the same effects as drinking on the job. OSH originally published a guide for employers on stress and fatigue in 1998. They are currently reviewing the guide with specific attention to stress and fatigue in two working situations.

The first situation is work that inherently contains situations or circumstances that are difficult to cope with. This may include healthcare workers, the police or social workers. The second situation is work that does not inherently contain stressful factors but is organized in such a way that coping may be difficult.

In the UK, the Health and Safety Executive (HSE) has published a comprehensive guide on preventing work-related stress. In the UK stress-related illness is responsible for the loss of 6.5 million working days each year, costing employers around £370 million and society as a whole as much as £3.75 billion. An estimated half a million people in Britain are suffering from work-related stress, anxiety or depression at levels that make them ill.

According to HSE's stress spokesperson, Elizabeth Gyngell, "Work-related stress is a huge occupational health problem facing Britain today, inflicting a heavy toll both in terms of financial cost and human suffering. We recognise that there is considerable pressure in the modern competitive work place, but there is a difference between the buzz people get from doing a busy job and staff simply being unable to cope with the strains placed upon them. A burnt-out workforce is an unproductive workforce and it is in no one's interests to find themselves in this situation."

The UK Health and Safety Commission has identified stress as one of its eight priority programs aimed at reducing accidents, injuries and ill-health in the workplace. (At the time of this writing, £ 1 = $ 1.47.)

– Rachel Michael, M.Sc., AEP

Reprinted from Ergonomics Today™. Used with permission.

EU Tackles Rising Costs of Workplace Stress

August 28, 2002

In July, the European Agency for Safety and Health at Work announced the first pan-European campaign to combat work-related stress. Some studies have linked work-related stress to increased numbers of musculoskeletal disorders (MSDs) as well as decreased productivity and quality.

Some facts about work related stress in the European Union:

- 28% of employees in the EU report they are affected by work-related stress;
- Women appear to suffer slightly more than men;
- Common causes include lack of job security and control and work overload;
- Over 50% of absenteeism has its roots in work-related stress;
- The problem is thought to cost the EU at least €20 billion a year in lost time and health costs.

(At the time of this writing, 1 EUR = 0.972259 USD.)

A major campaign to combat work-related stress – the second biggest occupational health problem in the European Union after back pain – was launched by the European Agency for Safety and Health at Work on 2 July 2002 at the European Parliament in Strasbourg. The initiative, which forms the heart of the Agency's annual European Week for Safety and Health at Work, will be the first EU-wide campaign to tackle work-related stress, a problem that is estimated to affect over 40 million employees in the EU.

According to a number of major studies, work-related stress affects nearly one-in-three

(28%) workers in the EU's 15 member states. Women report the highest levels, but for both sexes stress can be a problem in all sectors and at all levels of an organization. One of the most common causes is lack of control at work. 35% of employees, for instance, say they have no say in the order of their tasks and 55% claim no influence over how long they work. Monotony, tight deadlines (29% of staff claim to work regularly to these) and bullying are some of the other factors that enter the equation.

The human costs are significant. It is estimated that 16% of male and 22% of female cardiovascular diseases in the EU are due to work-related stress. Other diseases and conditions associated with this issue include musculoskeletal disorders and mental health problems.

The economic costs are equally large. Between 50% and 60% of absenteeism, for instance, has been related to work-related stress. Together with allied health costs, the total annual bill to the EU is estimated to be around €20 billion, not to mention loss of productivity.

"Stress is a growing problem with massive human and financial costs, but it is not one we have to accept," says Mrs. Anna Diamantopoulou, European Commissioner for Employment and Social Affairs. "The Commission is convinced that work-related stress can and must be prevented. However, we know that the only way it can be tackled effectively is by working together. That means involving everyone: from policy makers to researchers to social partners to the individual workers themselves. The Agency's campaign is an important step in this direction."

The campaign, which is being backed by all Member States, the European Commission and Parliament, trade unions and employers federations, will run until October and end with the European Week for Safety and Health at Work.

To support this drive, the Agency's campaign will raise awareness of the risks of work-related stress and suggest effective solutions through a range of media. This will include comprehensive information packs in all 11 EU languages, posters, leaflets and a multilingual website at http://osha.eu.int/ew2002. The Agency will also announce the winners of its European Good Practice awards for the organizations that have most successfully tackled work-related stress and some of its key triggers. Events will be organized in October across Europe to bring this message to life in all organizations, large and small, public and private.

- Edited by Rachel Michael, M.Sc., AEP

Reprinted from Ergonomics Today™. Used with permission.

Workplace Stress and Injury - the Latest Thinking

September 16, 2002

French researchers write that a comprehensive ergonomics program which focuses on both physical and psychosocial risk factors can significantly improve working conditions as well as save money for companies.

Researchers from the Laboratory of Biomechanics and Ergonomics and the National Institute for Research and Safety in France compiled the most recent data regarding stress and the occurrence of work related musculoskeletal disorders of the upper extremities (WRMSDue). The researchers point out that the role of stress and work related psychosocial factors in the development of WRMSDue is still poorly understood and there is no consensus on the epidemiological data. They do, however, propose that there is strong evidence that human responses to stress can present risk factors that may cause injury.

The researches focus on the physical reactions to stress and how they may increase the risk for injury. For instance, the researchers note that when a person experiences stress, the body releases certain chemicals. Some of these chemicals are pro-inflammatory and can lead to tendon inflammation. Others, called corticosteroids, can lead to oedema or swelling in the joints. Swelling and increased pressure in the joints is a risk factor for developing disorders such as Carpal Tunnel Syndrome. The researchers suppose that these factors mixed with poor ergonomics in the physical workplace lead to costly worker injury and illness. They support a comprehensive ergonomics program that looks at minimizing both physical risk factors such as high force or awkward postures and workplace stress.

The authors cite several studies including one reported by the European Foundation for the Improvement of Living and Working Conditions in which an industrial group addressed psychosocial factors in the workplace. The action reportedly brought not just financial, turn-over, absenteeism, productivity and other benefits, but also very significantly reduced the incidence of WRMSDue (255 cases a year in 1988 to 10 in 1994).

Stress is defined by the European Agency for Safety and Health at Work as "A psychological state which is part of and reflects a wider process of interaction between the person and their work environment ... stress may be experienced as a result of exposure to a wide range of work demands and, in turn, contribute to an equally wide range of heath outcomes."

In the U.S. the National Institute for Occupational Safety and Health (NIOSH) says that "the epidemiologic studies of upper extremity disorders suggest that certain psychosocial factors have a positive association with these disorders", but qualifies this with the assertion that "these factors, while statistically significant in some studies, generally have only modest strength."

Source: TUTB Newsletter. N. 19-20, September 2002
http://www.etuc.org/tutb/uk/pdf/2002-19p50-56.pdf

– Rachel Michael, M.Sc., AEP

Reprinted from Ergonomics Today™. Used with permission.

Guidelines Look to Regulate Workplace Stress

June 25, 2003

While OSHA drafts ergonomics guidelines for poultry and retail grocery industries to tackle musculoskeletal disorders (MSDs), government officials in the United Kingdom are focusing their efforts on just one possible contributor to work-related injuries: stress.

Developing draft processes for dealing with stress in the workplace, the UK's Health Safety Executive (HSE) released its Management Standards for Stress late last week. Working as an extension of a pre-existing pilot project on workplace stress, the goal of the draft standard, says the HSE, will be to help employers make the workplace less stressful for employees. Currently, the HSE estimates that 13.4 million workdays are lost annually in the UK because of stress-related illnesses.

Specifically, the HSE stated it developed the standards to address "stressors which affect the majority of employees in [the work] organization. In this way, the interventions [an employer makes] to meet the standards will have the greatest impact on the workforce as a whole. However, [employers] also need to address the concerns of individuals and small groups."

To develop the standards, the HSE used the assumption that 20 percent of the employees in the workforce are currently feeling "very" or "extremely" stressed because of work.

In the draft Management Standards, the HSE links workplace stress to six contributors: demands, control, support, role, relationship and change. Success will be achieved in an organization when 85 percent of employees within the organization state that demands, control and support are being managed well in the workplace and when 65 percent of employees state that role, relationship and change are being well managed.

The project was initially piloted in 24 companies who have responded with positive feedback.

The HSE also offers reasons why an employer should address work-related stress including an ethical argument that states work-related stress can cause illness and affect the quality of life of the employee, a legal argument reminding employers that sanctions can exist for employers if work-place stress is not addressed, and an economic argument reminding employers of the dollar amount that can be applied to workdays lost due to employee stress.

The pilot study of the draft Management Standards is expected to be finalized by this fall with a comment period running through next spring.

In the past, studies have linked continual stress to a higher risk of work-related injuries including MSDs. According to the American Psychological Association, small bouts of stress can be healthy, but continual stressful situations keep the body in "a constant state of activation," increasing the rate of wear to biological systems and ultimately making a person more susceptible to illness or injury.

Source: BBC.com

- Jeanie Croasmun

Reprinted from Ergonomics Today™. Used with permission.

It's the Work, Not the Hours, That'll Get You

August 25, 2003

Think long days are stressing out the workplace's front lines? A new study says that in actuality, it's the workplace and the work environment that builds stress in employees.

A study of nearly 3,000 American workers, presented at the annual meeting of the Academy of Management in Seattle, indicated that overall, workers in lower paying jobs, particularly service-oriented jobs, reported more job-related stress than did their white-collar counterparts, regardless of the number of hours the workers worked. The study reviewed a cross-section of America's workforce including full-time and part-time employees in a wide variety of job positions and fields, reported *Psychology Today*.

According to the report, regardless of the time spent working each day, workers who were less stressed were also more satisfied with their jobs. Additionally, jobs that offered employees more independence, more control over their work, supportive supervisors and fringe benefits were credited as causing less worker stress.

Aside from worker satisfaction and productivity, stress has also been linked to physical problems in workers, including an increased risk for work-related injuries and musculoskeletal disorders (MSDs).

Source: Psychology Today

– Jeanie Croasmun

Reprinted from Ergonomics Today™. Used with permission.

Improved Productivity Doesn't Have to Equal Increased Stress

September 10, 2003

With recent productivity jumps indicating that American businesses are getting more bang for their employee buck, it seems like the picture being painted by last quarter's 6.8 percent productivity increase would be a bright one. But more often than not in the recent weeks, those positive productivity improvements were turned into negatives: employee stress, job burnout, overworked workers.

A report earlier this month on MSNBC noted that Americans are reaching high work-related stress levels. The reason? Layoffs, pay cuts and cut-short vacation time means those employees who remain at a company have to pick up the slack of those who are gone. And Ronald Downey, Kansas State University Professor of Industrial and Occupational Therapy, concurred, telling MSNBC, "As the workforce has shrunk, people are overloaded and stress is the result. If the stress keeps on unending, then they're in trouble."

By trouble, Downey was referring to employees who take on more job responsibilities while losing control over work, racking up long work hours and trading in more of their free-time for work time. Stress, it has been found, can lead to physical health problems as well including an increased risk for developing musculoskeletal disorders (MSDs). According to the MSNBC report, the American Psychological Association estimates that absences due to psychological problems cost the workplace over $57 billion annually and health care costs are nearly 50 percent higher for workers reporting high levels of stress.

But productivity doesn't have to mean more stress on workers. By applying some basic ergonomics principals, workers can be more productive without being more stressed.

It simply comes down to understanding how people work and designing the workplace to best match those characteristics with business objectives. Is it easier to move a box from a shelf to a platform and back to a different shelf when packing it or could the entire process be done in one location? Is it faster for an attorney to dictate a letter that the assistant then listens to and retypes, edits and proofreads or could voice recognition software replace the dictation step?

Saving steps and increasing productivity and quality have long been goals of ergonomics, and these business goals should remain as valuable by-products when ergonomics is applied with a safety and health focus. As businesses look for new ways of solving the problems associated with workforce stress, ergonomics' productivity-enhancing abilities and its positive impact on a business' bottom line should be primary reasons to initiate ergonomic improvements.

Sources: Ergoweb Inc., MSNBC.com, ABCNews.com

- Jeanie Croasmun

3.7 References and Additional Information

Panero, J., and Zelnik, M. Human Dimension & Interior Space. New York: Watson-Guptill Publications, 1979.

Pheasant, Stephen. Bodyspace: Anthropometry, Ergonomics and Design. Philadelphia: Taylor & Francis, 1986.

4. Office Ergonomics Standards and Guidelines

4.1 Introduction

A standard is a requirement, rule or recommendation based on proven principles and practices. It is established by a recognized authority for the measurement of quantity, weight, extent, value or quality.

In addition to standards, there are several other types of product and environmental requirements, including laws, directives, guidelines and technology agreements. Guidelines usually are written and derive consensus from a small group. They usually do not carry the same level of authoritative weight as standards. Numerous categories and labels are applied to documents that are not standards. For the most part, a document that is not a standard is considered a guideline.

4.2 Voluntary Standards and Guidelines

4.2.1 International Organization for Standardization (ISO)

4.2.1.1 ISO 9241 (Ergonomic Requirements for Office Work With Visual Display Terminals) Overview

ISO 9241 establishes image quality requirements (performance specifications) for the design and evaluation of single- and multi-color computer monitors. At present, the recommendations are based on Latin, Cyrillic and Greek origin alphabetic characters, and Arabic numerals. Office tasks include such activities as data entry, text processing and interactive inquiry.

Part 1: General Introduction contains general information about the standard and provides an overview of each of the parts.

Part 2: Task Requirements discusses the enhancement of user interface efficiency and the well being of users by applying practical ergonomic knowledge to the design of VDT work tasks.

Part 3: Visual Display Requirements specifies requirements for visual displays and their images.

Part 4: Keyboard Requirements specifies the characteristics that determine the effectiveness in accepting keystrokes from a user.

Part 5: Workstation Layout and Postural Requirements specifies the design characteristics of workplaces in which VDTs are used.

Part 6: Work Environment Requirements specifies characteristics of the working environment in which VDTs are used.

Part 7: Display Requirements With Reflections describe how to maintain usable and acceptable VDT image quality by evaluating the reflection properties of a screen and the image quality of the screen over a range of typical office lighting conditions.

Part 8: Requirements for Displayed Colors states specifications for display color images, color measurement metrics and visual perception tests.

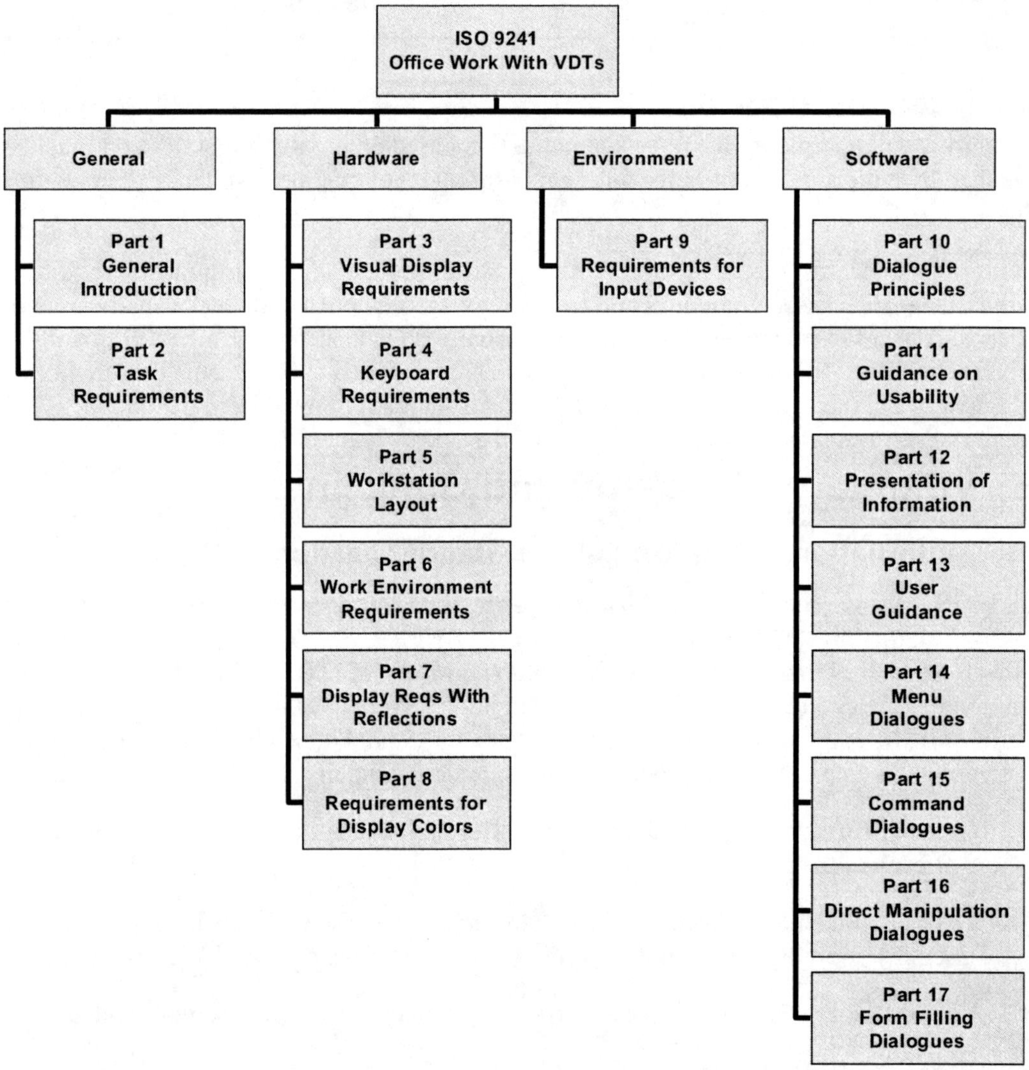

Organization of ISO 9241 by subject

Part 9: Requirements for Non-Keyboard Input Devices specifies requirements for the design and usability of input devices other than keyboards.

Part 10: Dialogue Principles specifies a set of high-level dialogue design principles for command languages, direct manipulation and form-based entries.

Part 11: Guidance on Usability explains the way in which the user, equipment, task and environment should be described-as part of the total system, and how usability can be specified and evaluated.

Part 12: Presentation of Information specifies requirements for the coding and formatting of information on computer screens.

Part 13: User Guidance specifies requirements and attributes to be considered in the design and evaluation of the software user interfaces.

Part 14: Menu Dialogues provides conditional requirements and recommendations for menus in user-computer dialogues.

Part 15: Command Dialogues provides conditional recommendations for common languages.

Part 16: Direct Manipulation Dialogues provides guidance on the design of manipulation dialogues in which the user directly acts upon object or object representations (icons) to be manipulated.

Part 17: Form Filling Dialogues provides recommendations for systems that use form filling interfaces (now commonly seen on the Internet).

Reprinted from Ergonomics Today™. Used with permission.

ISO Releases Software Accessibility Standard

June 27, 2003

Guidelines were released in June by the International Organization for Standardization (ISO) regarding technical specifications for designing software to be more accessible for the elderly and disabled as well as every user. Similar in scope to guidelines for website accessibility drafted by the World Wide Web Consortium (W3C), the ISO guidelines are "intended to provide a road map to software design that takes into account the varying physical and sensory capabilities of users to ensure that it can be used not only by people with visual, hearing, motor or cognitive disabilities, but also by the elderly, the temporarily disabled, and by those with no disability at all."

Tom Stewart, chair of the committee that developed the new guidelines, said the goal of the standard is to make computer systems more "usable, in particular, more effective, more efficient and more satisfying for all users. We want all users to be able to live and work in harmony with their computer systems," Stewart told UK internet legal news service, out-law.com.

The software accessibility standard is ISO/TS 16071, "Ergonomics of human-system interaction - Guidance on accessibility for human-computer interfaces."

Source: www.out-law.com

- Jeanie Croasmun

4.2.2 ANSI/HFES 100-1988 American National Standard for Human Factors Engineering of Visual Display Terminal Workstations

This standard was written through cooperation between the American National Standards Institute and The Human Factors Society. It specifies conditions for visual display terminals, associated furniture and the office environment. The standard was developed as a guideline for furniture manufacturers, not necessarily as a guide for office assessments and individual workstation recommendations. ANSI/HFES 100-1988 will be superceded by the BSR/HFES 100 document, described later in this chapter, if that document survives the ANSI consensus process. An overview of ANSI/HFES 100-1988 is presented here for historical and informational purposes only. Interested readers are encouraged to determine the status of the new BSR/HFES 100 and refer to the BIFMA and CSA standards, described later in this chapter, for more current guidelines.

4.2.2.1 Select Standard Office Workstation Dimension Recommendations

1. Height of the work surface that is supporting the keyboard should allow an angle range between the arm and forearm of 70 to 90 degrees for the 5th percentile female to the 95th percentile male: 23 to 28 inches.

2. Work surface width is dependent on the job requirements. Minimum width is elbow-to-elbow breadth of the 95th percentile male: 19.9 inches.

3. Viewing distance to the monitor is dependent on the size of the display characters and focusing capabilities of the user: minimum distance of 12 inches.

4. Thickness of the work surface should be approximately one inch.

5. Monitor height should be such that the topmost line of the display is not higher than the user's eyes.

6. Knee room height should be the distance from the floor to the highest point on the thigh or knee: a range from 20.2 inches for the 5th percentile female to 26.2 inches for the 95th percentile male for an adjustable work surface; 26.2 inches for a nonadjustable work surface.

7. Knee room width should be one inch larger than the thigh breadth of the 95th percentile female plus allowance for clothing: 20-inch minimum.

8. Knee/foot room depth under the work surface should be approximately 75 percent of the sitting buttock to popliteal distance plus the popliteal height times the sine of the angle between the vertical and the leg for the 95th percentile male: approximately 23.5 inches.

9. Seat height should allow the worker to have both feet on a stable support surface. It is dependent on the leg length of the 5th percentile female to the 95th percentile male, shoe heel height, angle of the leg while sitting, and the height/angle of the foot support: 16 to 20.5 inches.

10. Seat pan depth should allow the worker's back to be supported by the backrest without pressure against the backside of the leg. Minimum seat pan width should be the thigh width of the seated person. Seat pan depth should be 15 to 17 inches and seat pan width a minimum of 18.2 inches.
11. Seat pan angle should allow the angle between the leg and the thigh to be within a range of 60 to 100 degrees: 0 to 10 degrees backward slope.
12. Back rest design should be contoured to the shape of a lumbar curve and have dimensions of 6 to 9 inches high and 13 inches wide.
13. Backrest height should position the backrest to support the curve of the lumbar spine and be adjustable from 3 to 6 inches above the seat.
14. Backrest tilt should allow an angle between the thigh and torso of 90 to at least 100 degrees.
15. Angle between the backrest and seat should be 90 to 105 degrees.
16. The angle between the seatpan and the leg should be 60 to 100 degrees.
17. The height of the keyboard should allow a range of the angle between the arm and forearm of 70 to 90 degrees.

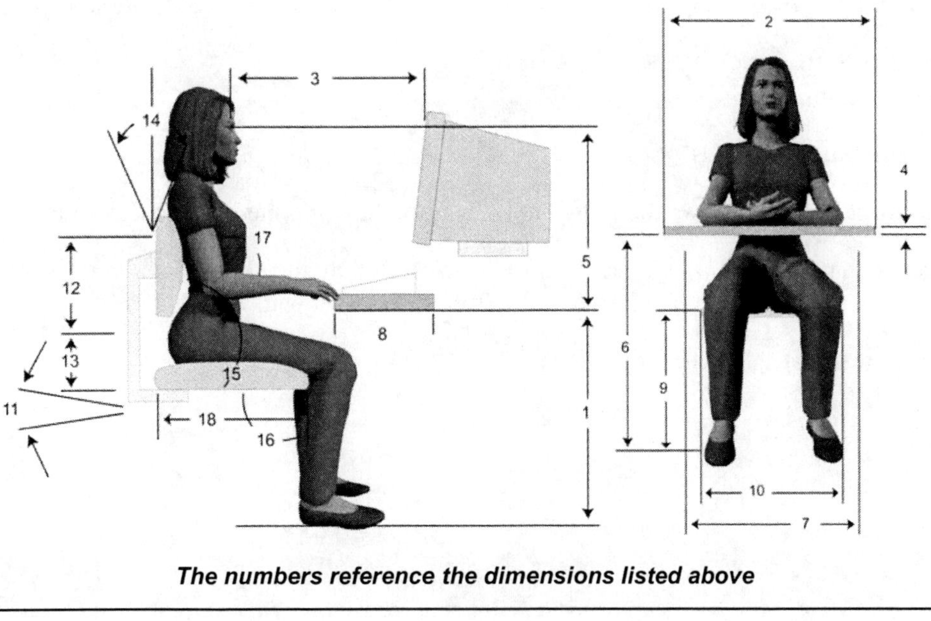

The numbers reference the dimensions listed above

4.2.2.2 Using a Laptop

Controlling the risk factors/risk factor modifiers will reduce the risk of injury while using this piece of equipment.

Suggestions include:

- Plug a regular sized keyboard into the laptop and use it for keyboarding.
- Plug a mouse or other cursor control into the laptop and use it instead of the laptop mousing device.
- Raise the screen such that the top of it is at eye level (some detach; with others, raise the whole unit).
- Use a docking station.

4.2.2.3 The Home Office

The risk conditions that exist in the regular office also exist in the home office. Considering some basic steps can abate them. The type/duration of tasks and room configuration influence the equipment needed in a home office. For example:

- Using a headset is a good idea if 4 hours a day are spent interacting on the telephone.
- The work surface should be large enough for your paper work or computer space needs.
- Equipment size/configuration should match the physical space of your home work area.

4.2.2.4 Equipment Guidelines: The Chair

Favorable chair features include:

- A minimum of five legs.
- A chair height that allows the thighs to be horizontal while the feet are flat on the floor.
- A seat pan depth that allows the back to be supported by the chair backrest while the back of the knee is comfortable relative to the front of the seat pan.
- A back rest that supports/fits the curvature of the lower back.
- Comfort.

4.2.2.5 Equipment Guidelines: The Computer Work Surface

Favorable work surface features include:

- If you plan to place the keyboard on the work surface, a good work surface height is approximately just below elbow height (you should be sitting with the feet flat on the floor, shoulder in neutral and the elbow at 90 degrees).
- Some work surfaces are adjustable in height. An adjustable height range from 25 to 30 inches would fit a large portion of the adult population.
- If you plan to place the keyboard on an adjustable height keyboard tray, the height of the work surface can have some variability. A vertical height between 27 to 31 inches will likely work well.
- The edges of the work surface should be rounded.

4.2.2.6 Equipment Guidelines: The Reading/Writing Work Surface

If you have to interact with a computer and perform a significant amount of paperwork, it is generally believed that a different work surface should be used for each activity. This is suggested because:

- A reading/writing work surface is usually best at a vertical height higher than a computer workstation.
- One work surface usually does not allow for enough space for both computing and paper work activities and can promote awkward postures.
- A favorable vertical height for the reading/writing work surface is one that feels comfortable to you when you are sitting with your feet on the floor and thighs horizontal to the floor. This usually ranges between 27 and 31 inches for most workers.
- The edges of the work surface should be rounded.

4.2.3 BIFMA: Ergonomics Guideline for VDT Furniture Used in Office Work Spaces

On July 20, 1999 the Ergonomics Subcommittee of the Business and Institutional Manufacturers Association (BIFMA International) voted to develop a guideline for furniture intended for computer use in the USA and Canada, by applying the measurable principles and design requirements of ISO 9241 parts 3 ("Visual display requirements") and 5 ("Ergonomic requirements for office work with visual display units").

BIFMA released the final version of this guideline in February 2002.

4.2.3.1 Working Postures

The Natick body dimensions are generally taken from upright postures, but there is no uniquely correct working posture that would fit any user for an extended period of time and/or accommodate every personal working habit. The dimensions given in this Guideline consider these generally upright postures only. Other dimensions may be acceptable if other postures are assumed.

Some typical working postures (based on seat pan angle) include:

- Rearward tilt (positive seat pan angle)
- Upright (near horizontal seat pan)
- Forward tilt (negative seat pan angle)
- Standing
- Alternating between sitting and standing

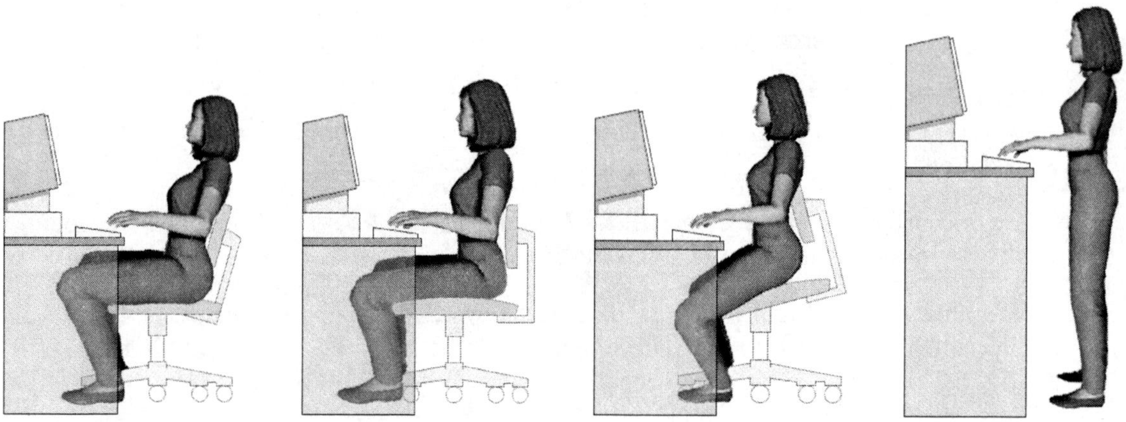

Rearward tilt ***Upright*** ***Forward tilt*** ***Standing***

All of these working postures are acceptable as long as the concept of fit is given proper consideration. ISO principles suggest that movement within and among these and other postures be encouraged.

4.2.3.2 The Work Chair

- The purpose of well-designed seating is to provide stable support that allows movement, comfort and task accomplishment.
- The purpose of good seating is to provide stable body support in a dynamic posture that is comfortable over a period of time, physiologically satisfactory and appropriate to the task or activity that is to be performed.
- The appropriate seat height for a user for sitting in the upright position is the popliteal height plus the thickness of footwear. Work chairs designed to accommodate a specified user population shall achieve fit for the range suitable for the intended user population. This range can be covered by applying the concept of fit.
- Within a selected range of adjustability, the seat height shall be user adjustable.

4.2.3.3 Relevant Body Dimension: Popliteal Height

Other considerations related to the work chair are seat depth, width and pan angle; back support height, width, lumbar support and adjustability; and armrest height and width.

4.2.3.4 Work Surfaces

- The main factors in determining appropriate workstation arrangements are seat and work surface, line-of-sight angle, work surface and keyboard height, knee clearance, forearm inclination and elbow height.
- Furniture, equipment and work environment may be designed for use in the seated or standing position and where sitting and standing alternate. Workstations need to be capable of supporting several tasks (screen viewing, keyboard input, non-keyboard input device usage, writing, etc.) and should therefore be designed with such functions in mind.
- The standing posture is recommended if it can alternate with a sitting position. This can be achieved if the workplace comprises either workstations or work surfaces for sitting and standing postures or an adjustable workstation that can accommodate the same person in the seated and standing position.
- Support surfaces for displays and input devices and associated equipment and materials should allow adequate clearance for the user's anthropometric characteristics and postural changes.
- For seated and standing work, sufficient vertical, horizontal and lateral clearance between the torso and lower limbs of users and workstation components is needed.
- The main considerations are for clearance for thighs, knees, lower legs and feet.

In the BIFMA guidelines, much time is spent in teaching how to correctly measure the relevant body dimensions. These guidelines are available through BIFMA International at http://www.bifma.org.

4.2.4 Canadian Standards Association: CSA-Z412

December 2000 saw the release of CSA's Z412 Guideline on Office Ergonomics. CSA is a nonprofit organization that oversees the development of standards and guidelines and also application of these standards through product certification, management systems registration and information products. The Z412 Guideline is a comprehensive document that includes design of jobs, work organization, office layout, environmental conditions and workstation design.

Z412 is structured around a nine step approach to a design process for obtaining an optimal office environment. (With the permission of Canadian Standards Association, material is reproduced from CSA Standard Z412-00, Guideline on Office Ergonomics, which is copyrighted by Canadian Standards Association, 178 Rexdale Blvd., Toronto, Ontario, M9W 1R3. While use of this material has been authorized, CSA shall not be responsible for the manner in which the information is presented, nor for any interpretations thereof.)

Step 1: Identify office design outcomes

This step defines office ergonomics and discusses various outcomes such as work performance, employee relations, and health and safety. Signs and symptoms of health and safety concerns such as musculoskeletal disorders and sick building syndrome are explained. The appraisal helps determine the nature of outcomes, early signs and symptoms, and risk factors of musculoskeletal disorders.

Step 2: Consider the office as a system and seek organizational involvement

This step discusses a systems approach to officer ergonomics. It also contains information about seeking the involvement of all parties within the organization who are affected by office design. The functions of an ergonomics committee for ongoing ergonomics efforts are described. The appraisal can be used to ensure you have considered the office as a system and achieved the involvement of all important parties.

Step 3: Identify workers' characteristics and seek their participation

This step helps the reader to understand the users of the office system. It contains information about anthropometry and aging as it affects office design. It discusses techniques for gathering user concerns. The importance of understanding worker needs and consulting with workers is explained. The appraisal evaluates the participation of workers in all steps in the office design process.

Step 4: Understand job demands, job design, and, and work organizations

This step provides basic information for conducting a task description or link analysis to understand the tasks and activities performed by the office worker. It also provides information about job design and organization of work, including issues of work scheduling and technological change. The appraisal helps identify if consideration has been given to optimal job design.

Step 5: Plan or consider the office layout and design of workrooms

This brief step describes issues related to general office layout, including the floor plan, work space, partitioning and storage areas. It also includes principles for design of office workrooms, including work heights, reaches, lifting/lowering, pushing/pulling and carrying. The appraisal evaluates office layout and the physical design of workrooms.

Step 6: Plan or consider the office environment
This section includes information on lighting, thermal environment, noise, air quality, whole-body vibration, electromagnetic frequencies and static electricity. In addition, consideration is given to cable management. The appraisal evaluates each of the environmental conditions.

Step 7: Design individual workstations (specify furniture, equipment, hardware, and layout)
This large step includes general workstation design principles and a discussion of various working postures. Performance goals and specifications are provided for furniture such as chairs, work surfaces, footrests, etc. Hardware features are discussed for monitors, keyboards, equipment (such as telephones). Layout and placement of items in a workstation are considered and several examples are provided. The appraisal summarizes all features of hardware, equipment and furniture design, as well as their layout and placement.

Step 8: Review the design
This step discusses how to present the design in an appropriate form (such as a mock-up) and ask for evaluations (user trials). The different people who may be involved and different methods for presenting and evaluating the design are discussed. The appraisal evaluates whether you have adequately reviewed the design.

Step 9: Provide education and training
This step provides basic information about adult education and the design of education and training sessions. Suggested topics are included for various participants in the office ergonomics education programs. The appraisal ensures education and training are appropriate.

This guideline is available from CSA.

Reprinted from The Ergonomics Report™, Vol. 1, No. 4, June 2002. Used with permission.

New Office Ergonomics Guidelines Contain Some Big Changes

Three new office ergonomics guidelines should take much of the confusion out of designing and setting up office computer workstations. The Canadian Standards Association (CSA) released "Guideline On Office Ergonomics" (CSA Standard Z412) in December 2000; the Business and Institutional Furniture Manufacturers Association (BIFMA) released "Ergonomics Guideline For VDT (Visual Display Terminal) Furniture Used In Office Work Spaces" in February 2002; and the Human Factors and Ergonomics Society (HFES) and the American National Standards Institute (ANSI) released "Human Factors Engineering of Computer Workstations" (BSR/HFES 100), which is a draft standard for trial use.

The BSR/HFES 100 document is actually the long overdue update to ANSI/HFES 100, which was originally published in 1998, and served as the primary guideline for North American furniture manufacturers until recently. (BSR is the ANSI designated acronym for "Board of Standards Review".) If a consensus is reached, and the standard is accepted and published, it will revert to ANSI/HFES 100. Presumably because ANSI/HFES failed to update their 1988 standard within the ANSI 10-year time frame requirement, BIFMA moved ahead with their own guideline.

There are some big differences between these new generation guidelines and the obsolete 1988 ANSI/HFES 100. For example, the original standard used an upright sitting posture as the reference posture that all furniture manufacturers should design to accommodate. This "90-90-90" posture, referring to the angles at the knees, trunk and elbows, was misinterpreted by many to be the only acceptable posture for computer work. This myth persists today, but these new guidelines and standards should expand the options and understanding for many.

As the new BSR/ANSI 100 committee chair Tom Albin put it, "I'd like to drive a stake through the heart of the idea that there's one correct position." In fact, both the BIFMA and BSR/HFES guidelines specifically recognize four basic reference postures for office computer workstations (see Figure 1):

1. Reclined sitting
2. Upright sitting
3. Declined sitting
4. Standing

Reference Postures

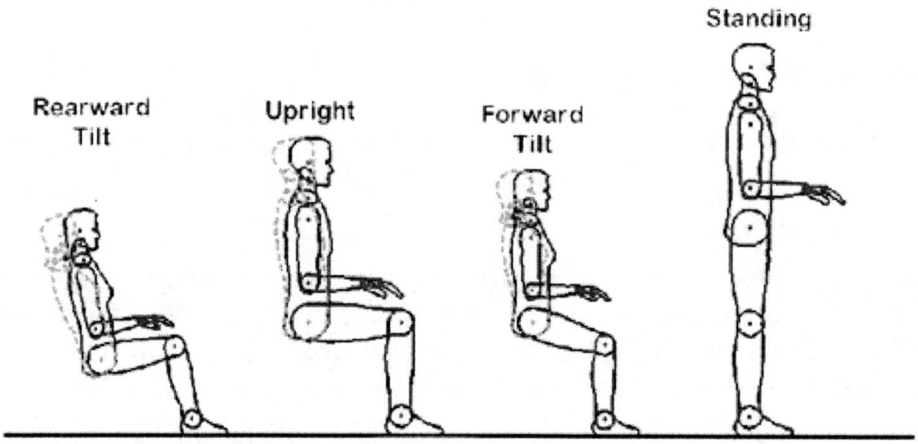

Office computing has changed dramatically since the 1980s when ANSI/HFES 100 was first published. Originally, keyboards were the only input devices discussed. The new version addresses a variety of input and display devices, including keyboards, mice, trackballs, joysticks, styli and light pens, tablets and overlays, and touch-sensitive panels.

Similarly, the display section has been expanded to cover both color and flat-panel displays.

[The BSR/HFES 100 document is available from HFES at a cost of US $50 for HFES members and US $85 for nonmembers. For details, go to http://www.hfes.org.]

The BIFMA guideline does not address computing equipment, but instead focuses on the concept of "fit" with respect to office furniture. A primary goal in its development was to provide an ISO 9241 guideline for North American markets.

[The BIFMA guideline "Ergonomics Guideline for VDT (Visual Display Terminal) Furniture Used in Office Work Spaces" is available at http://www.bifma.org or by contacting BIFMA International, 2680 Horizon Drive SE, Suite A-1, Grand Rapids, MI 49546.]

The BSR/HFES and BIFMA guidelines include sections specifically aimed at furniture and computer equipment manufacturers, but also have ample material aimed at end users in the office arena, including ergonomists, facility managers, health and safety personnel, and others that may be involved in the set-up of office computer workstations.

CSA-Z412 is a much more comprehensive document, and is aimed primarily at the end-user community as opposed to furniture and computer equipment manufacturers. It is also more process-oriented, and emphasizes a broad approach to include ergonomics in the design of office systems. The guideline even addresses telecommuting and lap top computing, and includes a section on material handling (lifting/lowering, pushing/pulling, carrying) in office settings. If you seek a broad, application-oriented guideline for office ergonomics, this comprehensive document might be the best choice among the three guides discussed in this article.

[CSA-Z412 is available at http://www.csa.ca or by contacting CSA International, 178 Rexdale Blvd., Toronto, Ontario, M9W 1R3, Canada.]

- Peter Budnick, Ph.D., CPE and Rachel Michael, M.Sc., AEP

4.3 Laws and Regulations

In addition to guidelines for the design and set up of office furniture and equipment, some states and localities have enacted ergonomics regulations. You should check in your region to learn whether there are such regulations in place. At the time of this writing, only the State of California has regulations in effect. Minnesota, Alaska and North Carolina have also considered enacting ergonomics regulations, though neither state, at the time of this writing, has done so.

Reprinted from The Ergonomics Report™, Vol. 1, No. 1, March 2002. Used with permission.

Regulating Workplace Ergonomics: State vs. Federal Approaches

The debate over workplace ergonomics rages on. It seems that no matter how hard regulatory opponents try to stop them, regulation supporters are bound and determined to put ergonomics rules into law. This type of political battle is not unique to ergonomics; it's a constant philosophical struggle, no matter what the underlying theme. In the US, there have been attempts to regulate ergonomics on local, state and federal levels. Depending on your philosophical bent, one or none of these approaches may be your wish. Some believe in decentralized approaches to government, where the more local the control, the better. Others believe that control should be centralized on the federal level. And still others believe that regulating things like ergonomics is wrong on all accounts, and should be left to free, peacefully cooperating individuals. Recognizing that there are going to be regulatory attempts, and assuming, for the sake of argument, that those attempts will eventually be successful, which would you prefer: state or federal level regulations?

The fact is, state level regulations are already in effect in California and Washington, and proposals are being considered in Minnesota and Alaska. Other states, such as North Carolina, have considered proposals, but withdrawn or defeated them, rejected the concept, or are waiting for the federal government to take the lead. If you've followed the workplace ergonomics debate, you know that the federal government first passed, then rescinded, a regulation focusing on controlling musculoskeletal disorders in general industry. This has left the door open for more states to move forward with their own regulations.

Critics of state level standards argue that each state will produce a different standard, with varying depth, breadth and enforcement. For businesses that operate in multiple states, this creates an expensive compliance and bureaucratic burden. Businesses that operate in one state only may experience more or less regulatory burden, depending on their state's approach, than their competitors in other states, creating unequal and unfair barriers or opportunities. From the labor perspective, workers in different states will receive unequal protections under the law. This effect is easy to see when comparing the existing California ergonomics regulation to the Washington law. California's law is approximately one page of text, while Washington's is much more specific and comprehensive. A federal law, supporters argue, would provide consistency and fairness across state lines.

Critics of federal regulations, on the other hand, argue that a single, "one-size-fits-all" standard is bound to be mediocre and less effective than regional approaches. A federal regulation only sets a minimally acceptable standard, and ignores the diverse cultural, business and economic conditions between states and regions.

In the end, there is no one right or wrong way for regulatory proponents to proceed. If a significant number of states move forward with ergonomics regulations, the federal government, out of necessity, will likely enact a law that does set minimum standards which all states must meet. Individual states are then welcome to exceed that minimum, but all must at least meet that minimum. And in the absence of a federal law, expect activists in more states to push regulations that set some minimum, if not some far reaching, ergonomics standards.

- Peter Budnick, Ph.D., CPE

4.3.1 State of California Ergonomics Standard

Since 1983 there has been organized activity within the State of California to develop ergonomics related standards to prevent work injuries. Initial efforts focused on video display terminals and were characterized by the adversarial perspectives of labor and business.

In 1993, the State passed Workers' Compensation Reform legislation that contained a mandate for the Occupational Safety and Health Standards Board to develop an ergonomics standard by January 1, 1995. The State of California OSHA wrote a proposed standard; however, the OSH Standards Board did not approve it.

January 1, 1995, arrived with no standard in place. The California Labor Federation sued the Standards Board for failing to meet the deadline. The judge ruled against the Standards Board and issued a court order for the Board to adopt a standard on ergonomics by December, 1996 (Occupational Safety and Health Standards Board, State of California, 1995).

Through 1995 and 1996, several versions of the ergonomics standards were written by the Standards Board. On November 14, 1996, the Standards Board adopted a measure that concentrated on repetitive motion injuries. The proposed standard was rejected by the state Office of Administrative Law (OAL) on January 2, 1997, due to poor clarity (Occupational Safety and Health Reporter).

The Standards Board readopted the standard on April 17, 1997 and resubmitted it to OAL for approval on May 16, 1997. The OAL approved the measure on June 3, 1997. The Standard became effective as a regulation on July 3, 1997.

California Ergonomics Regulation

TITLE 8

GENERAL INDUSTRY SAFETY ORDERS

SECTION 5110, ERGONOMICS

(a) Scope and application. This section shall apply to a job, process, operation where a repetitive motion injury (RMI) has occurred to more than one employee under the following conditions:

(1) Work related causation. The repetitive motion injuries (RMIs) were predominantly caused (i.e. 50% or more) by a repetitive job, process, or operation;

(2) Relationship between RMIs at the workplace. The employees incurring the RMIs were performing a job process, or operation of identical work activity. Identical work activity means that the employees were performing the same repetitive motion task, such as but not limited to word processing, assembly or, loading;

(3) Medical requirements. The RMIs were musculoskeletal injuries that a licensed physician objectively identified and diagnosed; and

(4)Time requirements. The RMIs were reported by the employees to the employer in the last 12 months but not before July 3, 1997.

(b) Program designed to minimize RMIs. Every employer subject to this section shall establish and implement a program designed to minimize RMIs. The program shall include a worksite evaluation, control of exposures which have caused RMIs and training of employees.

(1) Worksite evaluation. Each job, process, or operation of identical work activity covered by this section or a representative number of such jobs, processes, or operations of identical work activities shall be evaluated for exposures which have caused RMIs.

(2) Control of exposures which have caused RMIs. Any exposures that have caused RMIs shall, in a timely manner, be corrected or if not capable of being corrected have the exposures minimized to the extent feasible. The employer shall consider engineering controls, such as work station redesign, adjustable fixtures or tool redesign, and administrative controls, such as job rotation, work pacing or work breaks.

(3) Training. Employees shall be provided training that includes an explanation of:

(A) The employer's program;
(B) The exposures which have been associated with RMIs;
(C) The symptoms and consequences of injuries caused by repetitive motion;
(D) The importance of reporting symptoms and injuries to the employer; and
(E) Methods used by the employer to minimize RMIs.

(c) Satisfaction of an employer's obligation. Measures implemented by an employer under subsection (b)(1), (b)(2), or (b)(3) shall satisfy the employer's obligations under that respective subsection, unless it is shown that a measure known to but not taken by the employer is substantially certain to cause a greater reduction in such injuries and that this alternative measure would not impose additional unreasonable costs.

Note: Authority cited: Sections 142.3 and 6357. Labor Code. Reference: Sections 142.3 and 6357. Pulaski v.Occupational Safety & Health Stds. Bd. (1999) 75 Cal.App.4th 1315 [90 Cal. Rptr. 2d 54].

HISTORY

1. New article 106 (section 5110) and section filed 6-3-97; operative 7-3-97 (Register 97, No. 23).
2. Editorial correction of subsection (b)(1) (Register 97, No. 29).
3. Change without regulatory effect repealing subsection (a)(4) Exemption and amending Note filed 4-28-2000 pursuant to section 100, title 1, California Code of Regulations (Register 2000, No. 17).

Source: http://www.dir.ca.gov/title8/5110.html, September 18, 2003.

4.3.2 OSHA and Office Ergonomics

4.3.2.1 Introduction

Despite numerous and ongoing attempts to establish a federal ergonomics rule, the U.S. Occupational Safety and Health Administration, or OSHA, has failed to do so. At the time of this writing, OSHA is pursuing a guideline approach, part of which is a series of "eTools" available to the public on OSHA's web site (www.osha.gov). In August of 2003, OSHA released for public comment its Computer Workstations e-Tool. The e-Tool includes information on working postures, work practices and workstation components. It also includes a brief checklist on workstation setup. The public review period ends in May 2004.

According to OSHA, eTools are web-based products that provide guidance information for developing a comprehensive safety and health program. They include recommendations for good industry practice that often go beyond specific OSHA mandates.

Reprinted from Ergonomics Today™. Used with permission.

Bush Signs Repeal of Ergonomics Program Rule into Law

March 21, 2001

Tuesday marked the end of the present battle over ergonomics regulations as President Bush privately signed into law the repeal of OSHA's Ergonomics Program Standard. In a public statement the President wrote, "The safety and health of our nation's workforce is a priority for my administration. Together, we will pursue a comprehensive approach to ergonomics that addresses the concerns surrounding the ergonomics rule repealed today." The statement also said that Bush felt the rules imposed by his predecessor were "unduly burdensome and overly broad."

Several organizations have commented on this latest development. CNN.com reports that the AFL-CIO immediately issued the statement, "It's a sad day for injured workers." The Washington Post reported Martha G. Burk, chair of the National Council of Women's Organizations, as saying women suffer many ergonomic injuries from keyboard work and machine cleaning, and called the repeal "a slap in the face of women."

In contrast to these opinions, White House spokesman Ari Fleischer said Bush "believes that we can protect the health and safety of workers without passing a regulation that is terribly burdensome to the economy and to the small businesses on which their growth depends." He also said that Bush has directed Labor Secretary Elaine Chao to find ways to protect workers' health and safety without hurting businesses. Chao has said she would consider issuing a different ergonomics rule.

- Rachel Michael, M.Sc., AEP

Reprinted from Ergonomics Today™. Used with permission.

OSHA Announces Comprehensive Plan To Reduce Ergonomic Injuries

April 5, 2002

WASHINGTON - The Occupational Safety and Health Administration today unveiled a comprehensive plan designed to dramatically reduce ergonomic injuries through a combination of industry-targeted guidelines, tough enforcement measures, workplace outreach, advanced research and dedicated efforts to protect Hispanic and other immigrant workers.

"Our goal is to help workers by reducing ergonomic injuries in the shortest possible time frame," said Labor Secretary Elaine L. Chao. "This plan is a major improvement over the rejected old rule because it will prevent ergonomics injuries before they occur and reach a much larger number of at-risk workers."

Guidelines

Occupational Safety and Health Administrator John Henshaw said his agency will immediately begin work on developing industry and task-specific guidelines to reduce and prevent ergonomic injuries, often called musculoskeletal disorders (MSDs), that occur in the workplace. OSHA expects to begin releasing guidelines ready for application in selected industries this year. OSHA will also encourage other businesses and industries to immediately develop additional guidelines of their own.

Enforcement

The Department's ergonomics enforcement plan will crack down on bad actors by coordinating inspections with a legal strategy designed for successful prosecution. The Department will place special emphasis on industries with the sorts of serious ergonomics problems that OSHA and DOL attorneys have successfully addressed in prior 5(a)(1) or General Duty clause cases, including the Beverly Enterprises and Pepperidge Farm cases. For the first time, OSHA will have an enforcement plan designed from the start to target prosecutable ergonomic violations. Also for the first time, inspections will be coordinated with a legal strategy developed by DOL attorneys that is based on prior successful ergonomics cases and is designed to maximize successful prosecutions. And, OSHA will have special ergonomics inspection teams that will, from the earliest stages, work closely with DOL attorneys and experts to successfully bring prosecutions under the General Duty clause.

Compliance Assistance

The new ergonomics plan also calls for compliance assistance tools to help workplaces reduce and prevent ergonomics injuries. OSHA will provide specialized training and information on guidelines and the implementation of successful ergonomics programs. It will also administer targeted training grants, develop compliance assistance tools, forge partnerships and create a recognition program to highlight successful ergonomics injury reduction efforts.

Hispanic Outreach

As part of the Department of Labor's cross-agency commitment to protecting immigrant workers, especially those with limited English proficiency, the new ergonomics plan includes a specialized focus to help Hispanic and other immigrant workers, many of whom work in industries with high ergonomic hazard rates.

Ergonomics Research

The plan also includes the announcement of a national advisory committee; part of their task will be to advise OSHA on research gaps. In concert with the National Institute for Occupational Safety and Health, OSHA will stimulate and encourage needed research in this area.

"Bureau of Labor Statistics' data show that musculoskeletal disorders are already on the decline. This plan is designed to accelerate that decline as quickly as possible," said OSHA Administrator John Henshaw. "Thousands of employers are already working to reduce ergonomic risks without government mandates. We want to work with them to continuously improve workplace safety and health. We will go after the bad actors who refuse to take care of their workers."

The new plan was announced barely a year after Republicans and Democrats in Congress rejected the previous Administration's rule, which was developed over a period of eight years and was broadly denounced as being excessively burdensome and complicated. Over the course of the last year, the Department of Labor conducted three major public forums around the country and met with scores of stakeholders, collecting hundreds of sets of written comments and taking testimony from 100 speakers, including organized labor, workers, medical experts and businesses.

- Source: US Department of Labor

- Jeanie Croasmun

4.3.3 The Americans with Disabilities Act

4.3.3.1 Introduction

In 1990, Congress passed the Americans with Disabilities Act (ADA), a civil rights statute. The statute prohibits private employers, state and local governments, employment agencies and labor unions from discriminating against disabled individuals, who are otherwise qualified, in hiring, firing, promotion, compensation and training.

Application of the law requires comparing the physical abilities of the worker to the physical demands of the job essential functions. Ergonomics facilitates quantifying task requirements and provides the knowledge to design jobs to be more accommodating.

Passage of the ADA brought ergonomics into the civil rights arena. Under this law, the physical features of the workplace cannot be a barrier to employment opportunities for a qualified individual with a disability.

Ergonomists quantify task physical demands and can help define job essential functions that assist decision making in matching jobs with qualified individuals with disabilities.

Books and numerous newsletters offer information on the ADA. The purpose of this section is to provide a basic understanding of the law. The reference for all material presented in this section is A Technical Assistance Manual on the Employment Provisions (Title I) of the Americans with Disabilities Act.

4.3.3.2 ADA Overview

The Act:

- Protects qualified individuals from employment discrimination.
- Became effective July 26, 1992, for employers with 25 or more workers and July 26, 1994, for employers with 15 or more workers.
- Is enforced by the U. S. Department of Justice, Civil Rights Division, Disability Rights Section; remedies that may be required from a charge filed by a worker with the agency include:
    - Compensatory and punitive damages;
    - Back pay;
    - Front pay;
    - Attorney's fees;
    - As well as reasonable accommodation, reinstatement and job offers.

This requires the affected employer to consider the following issues:

- Who is a disabled person?
- If a disabled person applies for a position within the company, does the person have to be hired?
- What process changes and work accommodations are required?
- What are company obligations to an employee who becomes disabled due to an on the job injury?

This is being further defined as complaints are made to the EEOC and the courts make decisions.

4.3.3.3 ADA Law Definitions

The Law

The ADA is a federal law that prohibits an employer to discriminate against qualified individuals with disabilities in all aspects of the employment process.

A Qualified Individual
A person who:

- Satisfies the requisite skill, experience, education and other job-related requirements of the employment position such individual holds or desires;
- And who, with or without reasonable accommodation, can perform the essential functions of such position.

Reasonable Accommodation
Reasonable accommodation is any modification or adjustment to a job, an employment practice or the work environment that makes it possible for an individual with a disability to enjoy an equal employment opportunity.

Examples of reasonable accommodations include:

- Making facilities accessible and usable.
- Restructuring jobs.
- Modifying work schedules.
- Providing flexible leave policies.
- Acquiring or modifying equipment and devices.
- Adjusting and modifying examinations, training materials and policies.

Legal obligation to reasonable accommodation is limited by the imposition of undue hardship on business operation. This requires significant difficulty or expense in relation to the size of the employer, the resources available and the nature of the operation. The concept of undue hardship includes any action that is:

- Unduly costly;
- Extensive;
- Substantial;
- Disruptive;
- Or would fundamentally alter the nature or operations of the business.

Essential Functions
The essential functions are the basic activities that are required to perform a job.

Criteria for a function to be considered essential include:

- The position exists to perform the function.

Example: A person is hired to proofread documents. The ability to proofread accurately is an essential function, because it is the reason this position exists.

- There are a limited number of other employees available to perform the function or among whom the function can be distributed.

Example: It may be an essential function for a file clerk to answer the telephone if there are only three employees in a very busy office and each employee has to perform many different tasks.

- A function is highly specialized, and the person in position is hired for special expertise or ability to perform it.

Example: A sales position is open for a company doing business primarily in Japan. Knowing the Japanese language is an essential function of the job.

An Individual With a Disability
A person who:

- Has a physical or mental impairment.
    - A physiological or mental disorder that compromises the normal working of the body such as leg amputee, epilepsy, dyslexia or contagious disease (HIV).
    - Impairment does not include height or weight within normal range, pregnancy or inability to read due to dropping out of school.
- Has a record of such an impairment.
    - An individual may have a history of an impairment that currently may or may not effect them such as heart disease, cancer or rehabilitated drug user, or
- Is regarded as having such impairment.
    - An individual who is perceived as having an impairment such as high blood pressure that is under control or facial scar on a person with high public contact.

That:

- Substantially limits
    - An individual must be unable to perform, or be significantly limited in the ability to perform, an activity compared to an average person in the general population.

- One or more major life activities
  - Activities that an average person can perform with little or no difficulty such as walling, speaking, breathing, seeing, performing manual tasks, learning, sitting.
  - Standing, lifting and reading.

Direct Threat Exclusion

The employer may require as a qualification standard that an individual not pose a direct threat to the health or safety to the individual or others.

The employer must show:

- There is a significant risk of substantial harm.
- That a specific risk has been identified.
- It must be a current risk, not one that is speculative or remote.
- The assessment of the risk must be based on objective medical or other factual evidence regarding a particular individual.
- Seasonable accommodation can eliminate or reduce the concerns below the level of a direct threat.

4.3.3.4 Application of ADA

Employers cannot discriminate against individuals with disabilities in regard to any employment practices or terms, conditions and privileges of employment.

This includes:

- Application
- Testing
- Hiring
- Assignments
- Evaluation
- Disciplinary actions
- Training
- Promotion
- Medical examinations
- Layoff/recall

- Termination
- Compensation
- Leave
- Benefits

4.3.3.5 Decision Making

The following sequence is offered as a possible decision making sequence for application of ADA. This is not legal advice. If you have a specific situation with a worker or potential worker, seek appropriate legal counsel.

Question 1: Does the individual have requisite skills?

If NO: No job offer (be ready to defend decision by ADA rules).

If YES: Individual is classified as "otherwise qualified".

Question 2: Can the individual perform essential job functions?

If NO or DO NOT KNOW: Post offer/pre-employment medical exam.

If YES: Job offer can be made without concern of accommodation.

Question 3: Is there a disability?

If NO: Job accommodation is not required. No job offer.

If YES: Accommodation evaluation.

Question 4: Can reasonable accommodations be made so the individual can perform job essential functions?

If NO: This has to be due to undue hardship. No job offer.

If YES: "Direct threat" analysis.

Question 5: Is there a "direct threat"?

If NO: Job offer can be made. The individual is a qualified individual with a disability who can perform the essential functions of the job with reasonable accommodation without creating a direct threat.

If YES: No job offer.

4.3.3.6 Job Analysis

The ADA does not require that an employer conduct a job analysis to identify the essential functions of a job. However, to comply with ADA regulations, it is wise to know some specific job requirements. These include:

- ♦ Define requisite skills.
    - Identify the necessary education, experience and licenses required to hold the position.
- ♦ Define essential job functions.
    - Identify the purpose of the job (results or outcome) and the importance of actual job functions in achieving the purpose.

Purpose Example: Move heavy packages from the dock to a storage room as opposed to lift and carry heavy packages from the dock to the storage room.

Importance Example: The frequency, amount of time and the consequences of not performing a function.

- ♦ Define the manner by which a job is performed.
    - The manner by which a job is performed can be described. However, it is erroneous to conclude it is the only way a job can be done.

An ergonomist's skills are helpful in identifying the purpose of a job and quantifying job functions such as:

- ♦ Amount of force required to push a pedal/lever
- ♦ Weight of tool/load
- ♦ Reach distance
- ♦ Frequency of widget production
- ♦ Length of time task is performed

4.3.3.7 Actions That Constitute Discrimination

The following situations have been defined by EEOC as examples of violations of the ADA law:

- ♦ Refusing to make reasonable accommodation to the known physical or mental limitations of a qualified applicant or employee with a disability, unless the accommodation would pose an undue hardship on the business.
- ♦ Using qualification standards, employment tests or other selection criteria that screen out or tend to screen out an individual with a disability unless they are job-related and necessary for the business.
- ♦ Failing to use employment tests in the most effective manner to measure actual abilities. Tests must accurately reflect the skills, aptitude or other factors being measured, and not the impaired sensory, manual or speaking skills of an employee or applicant with a disability (unless those are the skills the test is designed to measure).

The U.S. Court system has begun to interpret the ADA and set precedence by deciding cases. One decision in particular is worth noting here (refer to the following articles).

Reprinted from Ergonomics Today™. Used with permission.

Supreme Court Releases Decision Involving ADA and Carpal Tunnel Syndrome

January 8, 2002

The Supreme Court of the United States has rendered an opinion that sets important precedence for the interpretation of the Americans with Disabilities Act (ADA). The decision helps clarify the meaning of "disability" and key ADA terminology including "substantially limit" and "major life activity." In a nutshell, the Court decided in favor of Toyota, and suggests that a disability must be evaluated on an individual, case-by-case basis; that a mere diagnosis of impairment is not sufficient evidence under the test of substantially limiting major life activities; that an impairment's impact must be permanent or long-term; that manual tasks in question must be central to daily life; and that impairments that limit only work activities do not meet the definition of "disability" in the Act.

In *Toyota v. Williams (No. 00-1089)*, Plaintiff Ella Williams, claiming to be disabled from performing her automobile assembly line job by carpal tunnel syndrome and related impairments, sued her former employer, Toyota, for failing to provide her with a reasonable accommodation as required by the ADA. A lower court had previously granted Toyota summary judgment, holding that Williams impairment did not qualify as a "disability" under the ADA because it had not "substantially limit[ed]" any "major life activit[y]," and that there was no evidence that she had had a record of a substantially limiting impairment or that Toyota had regarded her as having such an impairment.

The Sixth Circuit Court, who found that Williams's impairments substantially limited her in the major life activity of performing manual tasks, reversed that decision. In order to demonstrate that she was so limited, according to that court, Williams had to show that her manual disability involved a "class" of manual activities affecting the ability to perform tasks at work. She satisfied this test, said that court, because her ailments prevented her from doing the tasks associated with certain types of manual jobs that require the gripping of tools and repetitive work with hands and arms extended at or above shoulder levels for extended periods of time. In reaching that conclusion, Sixth Circuit found that evidence that Williams could tend to her personal hygiene and carry out personal or household chores did not affect a determination that her impairments substantially limited her ability to perform the range of manual tasks associated with an assembly line job.

The unanimous Supreme Court decision, delivered January 8, 2002, by Justice O'Conner, holds that the Sixth Circuit did not apply the proper standard in determining that Williams was disabled under the ADA because it analyzed only a limited class of manual tasks and failed to ask whether her impairments prevented or restricted her from performing tasks that are of central importance to most people's daily lives. The Court interpreted the word "substantially," as used in "substantially limits" in the ADA, as suggesting "considerable" or "to a large degree," and thus clearly precludes impairments that interfere in only a minor way with performing manual tasks. Further, because "major" means important, "major life activities" refers to those activities that are of central importance to daily life. Therefore, manual tasks in question must be central to daily life. To be substantially limited in the specific major life activity of performing manual tasks, therefore, an individual must have an impairment that prevents or severely restricts the individual from doing activities that are of central importance to most people's daily lives. The impairment's impact must also be permanent or long-term.

The Court holds that it is insufficient for individuals attempting to prove disability status under this test to merely submit evidence of a medical diagnosis of an impairment. Instead, the ADA requires them to offer evidence that the extent of the limitation caused by their impairment in terms of their own experience is substantial. That the ADA defines "disability" "with respect to an individual," makes clear that Congress intended the existence of a disability to be determined in a case-by-case manner. An individualized assessment of the effect of an impairment is particularly necessary when the impairment is something like carpal tunnel syndrome, in which symptoms vary widely from person to person.

The Court believes the Sixth Circuit erred in suggesting that in order to prove a substantial limitation in the major life activity of performing manual tasks, a plaintiff must show that her manual disability involves a "class" of manual activities, and that those activities affect the ability to perform tasks at work. The Sixth Circuit's analysis erroneously focused on Williams' inability to perform manual tasks associated only with her job. Rather, the central inquiry must be whether she is unable to perform the variety of tasks central to most people's daily lives. Also without support is the Sixth Circuit's assertion that the question whether an impairment constitutes a disability is to be answered only by analyzing the impairment's effect in the workplace. That the Act's "disability" definition applies not only to the portion of the ADA dealing with employment, but also to the other provisions dealing with public transportation and public accommodations, demonstrates that the definition is intended to cover individuals with disabling impairments, regardless of whether they have any connection to a workplace. Moreover, because the manual tasks unique to any particular job are not necessarily important parts of most people's lives, occupation-specific tasks may have only limited relevance to the manual task inquiry. In this case, repetitive work with hands and arms extended at or above shoulder levels for extended periods, the manual task on which the Sixth Circuit relied, is not an important part of most people's daily lives. Household chores, bathing and brushing one's teeth, in contrast, are among the types of manual tasks of central importance to people's daily lives, so the Sixth Circuit should not have disregarded Williams's ability to do these activities.

Source: Legal Information Institute, http://supct.law.cornell.edu/supct/html/00-1089.ZS.html

– Peter Budnick, Ph.D., CPE

Reprinted from Ergonomics Today™. Used with permission.

Interpreting Toyota vs. Williams: Does it Impact Ergonomics?

January 14, 2002

On January 8th, 2002, the U.S. Supreme Court ruled in favor of Toyota when it decided that former employee Ella Williams did not qualify as "disabled" under the Americans With Disabilities Act (ADA). A few have since speculated that the Court's decision dealt a blow to occupational ergonomics, but a close look at the ruling shows that is not true, and may in fact have a positive influence with respect to ergonomics in the workplace.

The only link to occupational ergonomics in the case was that Williams purportedly suffered from Carpal Tunnel Syndrome (CTS) and related ailments that resulted from her duties as an assembly worker in a Toyota Manufacturing plant. Whether or not her condition was caused by her work was not at issue, and it appears that the Court accepted her impairments and consequent physical limitations as fact. That is, Toyota did not ask the Court to decide on an issue of ergonomics; they did ask the Court whether her impairments qualified Williams as "disabled" under the ADA.

In essence, the Court responded that an impairment, regardless of its name or medical diagnosis, could be classified as a disability only if it limits a person's ability to perform "major life activities." The Court suggested that activities such as household chores, bathing and brushing one's teeth, for example, are of central importance to most people's daily lives. Performing a certain set of tasks on an assembly line, on the other hand, is not necessarily important to the daily lives of the broader American population. Presumably, had Williams impairments restricted her from household chores, bathing, etc., the Court would have defined her as disabled under the ADA.

Labor activists have responded that one's work life is a "major life activity" for most people, and that impairments that limit one's ability to work should be covered under the disability definition. The difference between this perspective and the Court's opinion seems to be one of an individual vs. population comparison. Some activists feel that the test should be whether an individual is disabled with respect to his or her own daily life, whereas the Court states that the test must be against the entire population. Williams couldn't do the specific tasks in her specific job, but she could still perform what most people do in their daily life activities. So, while the Court clearly stated that each individual must be evaluated on a case-by-case basis, the evaluation must compare him or her to the population, not only to themselves, in the determination of disability.

The Court also said an impairment must be long-term to qualify as a disability. Since the severity of impairments such as CTS may change over time, or may even heal completely, a mere diagnosis is not enough to prove disability.

Prior to this decision, business was concerned that if an employee had any impairment that restricted any work activities, the employee might have the makings of an ADA case. The U.S. Chamber of Commerce hailed the decision, saying "Expanding the ADA from covering workers with disabilities to include workers who cannot perform a particular job function would have exploded the cost of doing business, and dramatically increased all employers' risk for lawsuits from disgruntled workers."

The American Association of People with Disabilities (AAPD) criticized the decision. "These cases are about whether victims of disability discrimination get their day in court. Ignoring Congress's explicit use of an inclusive definition of 'disability' in the ADA, the

Supreme Court today and in earlier decisions has gone out off its way to narrow the definition of 'disability' and thereby limit access to the courts for victims of discrimination."

Some experts believe the ruling will have less impact than most observers expect. For instance, pointing to the differences between state and federal workplace laws, Fred Main, Senior Vice President of the California Chamber of Commerce, believes the Supreme Court decision will have little effect in his state. "As in many issues of employment law, California's courts are a more favorable venue than the federal courts for workers," he said. Claudia Senter, a staff attorney at the Legal Aid Society's Employment Law Center in San Francisco, notes that California law only a requires a "limitation" to a "life activity," as opposed to the stricter ADA language "substantially limits" and "major life activities." Therefore, California and other states may have a broader definition of disability in their state laws, and employees with disability claims will likely sue in state rather than federal courts as a result.

Williams vs. Toyota was not a case questioning ergonomics, nor was it a case questioning Carpal Tunnel Syndrome or its potential relationship to work. In fact, the impairments and potential work-relatedness associated with the disorder appear to have been taken for granted by the Court. The case really focused on the question of who is disabled and who is not, within the confines of the Americans With Disabilities Act. With this decision the Federal Court has narrowed and clarified that definition, making it more exclusive than many previously interpreted. State laws may have more inclusive definitions, and if so, expect to see certain disability cases fought in state courts rather than federal.

- Peter Budnick, Ph.D., CPE

5. Applying Ergonomics in the Office

5.1 Strategies for Improvement

There are many positive goals to strive for when applying ergonomics in an office environment, including:

- Reduced error rates and increased quality.
- Increased productivity.
- Reduced absenteeism and employee turnover.
- Improved morale.
- Reduced injuries and discomfort.

Each of these has measurable costs and opportunities for improvement. Often, they are complementary goals, and a system that improves one of these factors will improve others.

5.2 Some Basic Principles

5.2.1 There Is No "One Size Fits All"

Clothing comes in many sizes and styles because people come in many different sizes and exhibit different tastes and desires. Clothing also differs based on the environment in which it will be worn or the tasks that will be performed while wearing it.

For example, it's difficult to argue that it's unwise to wear shorts in a winter climate. Likewise, it's poor planning to place a computer in a location that does not fit the particular user (or users) performing specific tasks. Matching equipment to operators and their tasks, like matching clothing to people, environments and tasks, is necessary to achieve productivity, quality and safety goals.

5.2.2 There Is No "Best Posture" or "Best Way to Sit"

Conventional wisdom ("Sit up straight!") and a common misinterpretation of ANSI/HFES 100-1988 results in a persistent but misguided belief that there is one "proper" posture or one "proper way to sit" for all people, all the time, when performing computer tasks. In fact, this recommendation is likely to cause more harm than good.

5.2.3 "The Next Posture Is the Best Posture"

The original source of this statement is unknown, but it has become a popular phrase when training or when responding to questions like, "What's the best way to sit?" The static nature of office work is a root cause of discomfort, pain and injury. Frequent changes in posture are highly advised to encourage muscle contractions and consequent blood flow.

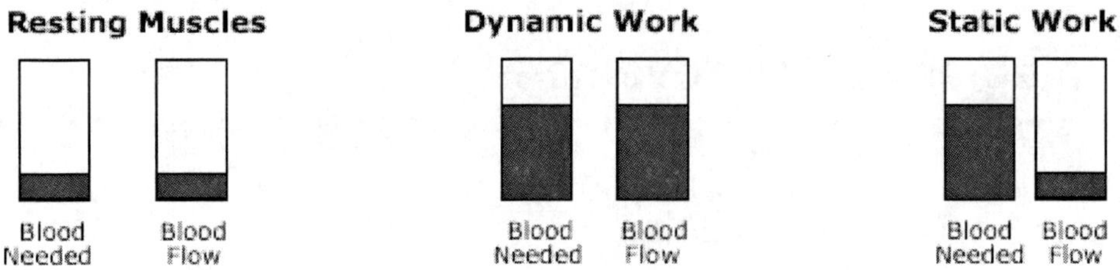

Blood flow is important to supplying nutrients to body tissues and removing metabolic byproducts. Static work can result in reduced blood flow.

Furniture that somehow encourages movement through design is referred to as "dynamic seating" or "passive ergonomics". It may also be possible to design work processes that require some level of movement, such as moving away from a computer to perform a filing task, or even intentionally locating some work tools at a distance that requires the operator to occasionally change postures in order to complete job requirements. Other approaches include scheduled breaks, frequent micro-breaks or stretching breaks. However, administrative methods like these, as opposed to designs that automatically encourage more movement, are likely to be less effective because they rely on human behavior and consequent management enforcement. Further, especially when following research recommendations of up to 10 minutes of rest/break per hour, administrative techniques can meet with management and employee resistance.

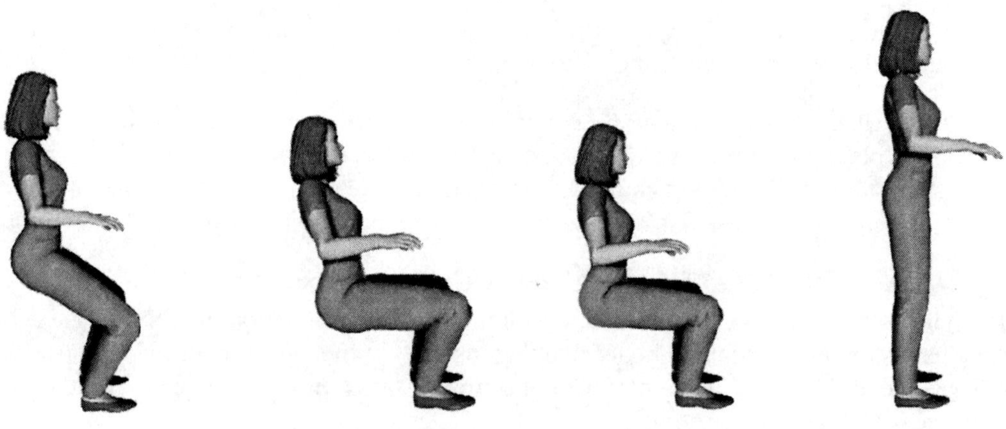

Four reference postures used by furniture designers to accommodate a variety of working postures.

5.2.4 The Hands and Eyes Drive Posture

It's common to see computer operators hunched or leaning forward, taking their back away from the chair backrest. And it's just as common to place the blame on the individual operator, rather than looking for the root cause of such postures.

Just like the location of a conveyor belt in a factory or the location of the tools and components on an assembly line define the postures of workers in those environments, the location of the keyboard, mouse and monitor drive the postures of office workers. In general, if a workstation allows the worker to comfortably reach necessary tools and devices while sitting in a well-supported posture, he or she will. On the other hand, if something is interfering with the ability to sit in a comfortable and productive posture, he or she won't.

5.2.4.1 Eliminate Barriers That Force Workers Away from Their Work

For example, armrests at the same height as a work surface or keyboard tray often keep the operator from being able to move the chair closer to the work. CPUs, files and personal items stored under a desk can have the same effect of forcing a worker away from their work.

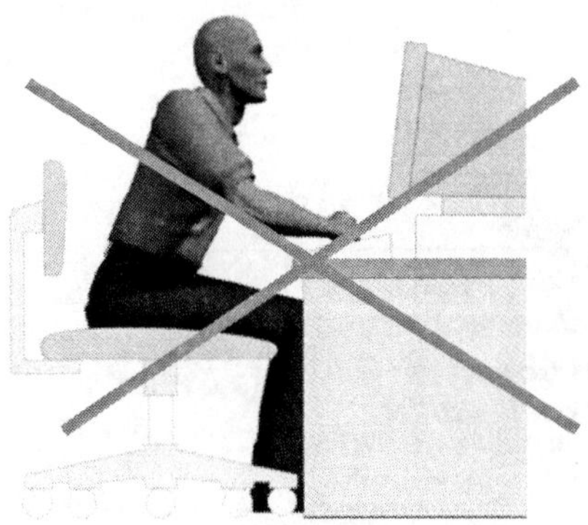

Obstructions can force the worker too far away from his or her work, resulting in a forward leaning posture.

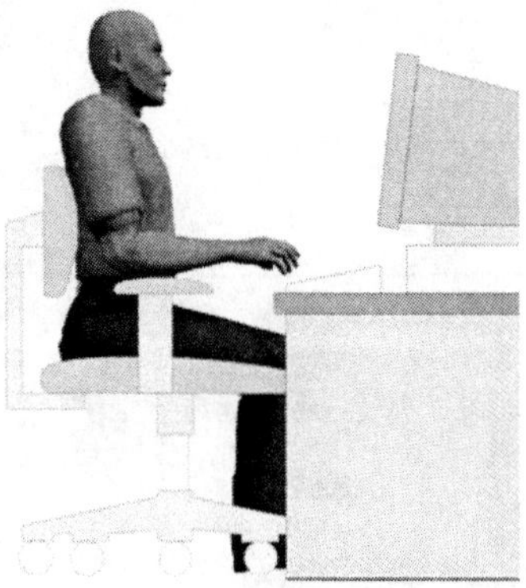

A worker should be able to sit comfortably and access his or her work without unnecessary reaching or bending forward.

Some people prefer chairs that support an open hip posture that allows them to sit closer to their work while still sitting in a supported posture. This is one example that demonstrates why a single chair design is typically not appropriate for each and every worker.

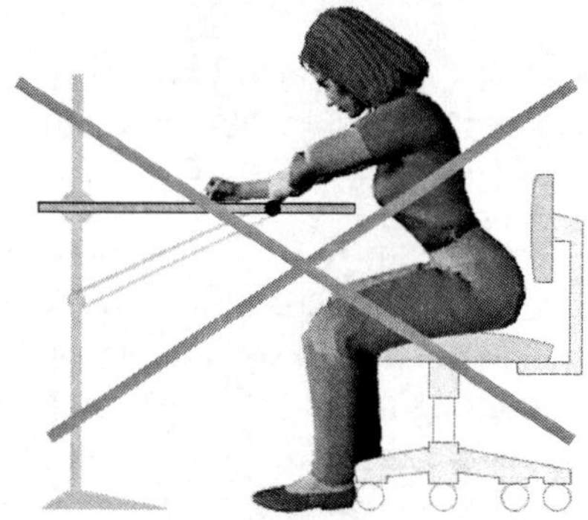

This workstation design, combined with the task, forces the user into an awkward posture.

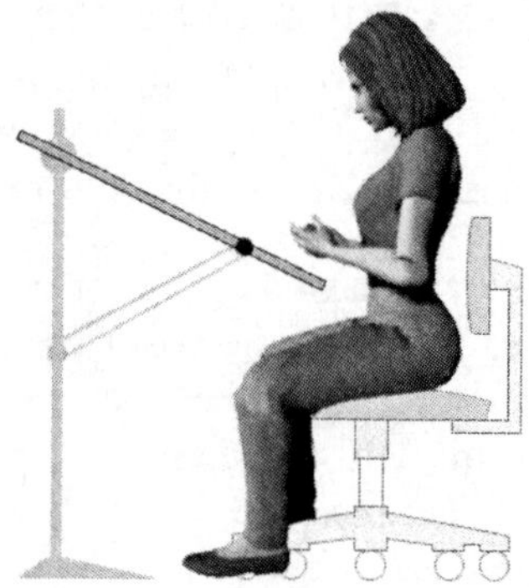

This workstation design allows the user to sit in a less awkward posture while performing the same task.

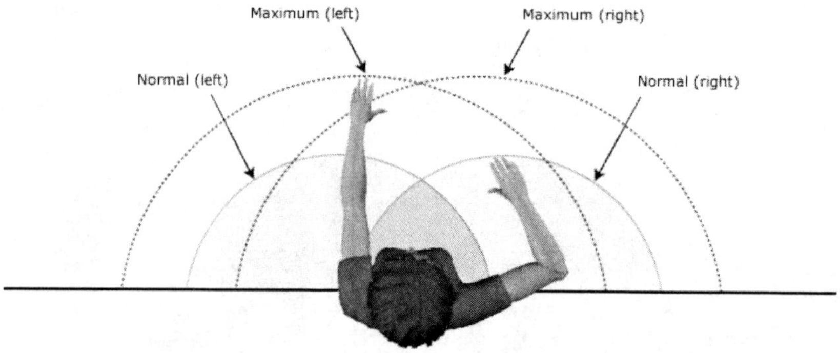

Commonly used equipment should be located within a comfortable reach zone for each worker.

5.2.4.2 Monitor Placement

Goals: Minimize eye, upper back and neck strain; reduce visual fatigue and errors.

Method: A rule of thumb for monitor placement, which appears to have little scientific basis, is that it should be at approximately arms length from the viewer, with the top of the monitor approximately at or below seated eye height. There is nothing inherently wrong with such a recommendation, but it should under no circumstances be considered a requirement. In fact, monitor placement, like all other component locations, is a function of the task, equipment and software characteristics, and operator preferences and visual capabilities.

In theory, there are two human eye visual characteristics that determine where "the best" monitor placement is for any individual (Ankrum, 1996):

- Accommodation, and
- Convergence

Accommodation occurs as the eye's lens capsule changes shape to focus at a particular distance. Convergence refers to the inward turning of each eye when viewing at close distances (for example, when you try to view your finger as you move it from a full reach to the tip of your nose). In practice, physical barriers like walls and cubicles, furniture and equipment dimensions, monitor quality, lighting and glare, eyeglasses and software design affect that "best viewing distance".

Reprinted from The Ergonomics Report™, Vol. 2, No. 11, November 2003. Used with permission.

Vision Tips For Older Workers

Want to make sure everyone in the workplace can actually see that they're on the same page? Dr. Jeffrey Anshel offers the following suggestions:

- **Lower the monitor.** According to Anshel, whether we're looking at a computer screen or a newspaper, when we read, our eyes turn in and turn down.
- **Watch out for glare.** Anshel recommends angling the monitor back somewhat to try to make the face of the monitor perpendicular to glare.
- **Address the lighting.** Most offices are designed for paper, says Anshel, indicating that paper, unlike computer monitors, isn't back-lit and needs illumination. But when computers and paper share the same space, consideration needs to be made for the fact that monitors are also a light source.
- **Adjust the brightness.** Anshel indicates that, while monitors might look snazzier in the showroom with their brightness controls set all the way to high, lighting in the workplace is different. Plus if natural light is a factor, a comfortable brightness setting might change for a worker throughout the day.
- **Older workers normally need more light.** Anshel also notes that workers with cataracts tend to work most effectively with white letters on a black background.

- Jeanie Croasmun

5.2.4.3 Keyboard Placement

Goal: Place the keyboard so that the operator can type efficiently, with minimal errors and minimal discomfort in the hands, arms, shoulder, neck and back.

Method: In theory, the keyboard should be located relative to the user such that the arms are not reaching forward or to the side (especially without support), the wrist angles are neutral and the keyboard is stable. In practice, workstation dimensions and layouts, combined with individual differences in preference and typing skill, result in awkward postures and less productive, more error prone arrangements. A common set-up that leads to awkward postures places the keyboard in a different plane from the monitor (below left). Utilizing an arrangement that keeps the monitor and keyboard in the same plane is preferable (below right).

An awkward posture that places the monitor and keyboard in a different plane.

A better posture that places the monitor and keyboard in the same plane.

5.2.4.4 Mouse/Pointing Device Location

Goal: Locate the pointing device such that the operator performs necessary clicks and movements accurately and with minimal discomfort.

Method: In theory, the pointing device should be located as close as possible, and at the same height and in the same plane, as the keyboard. In practice, many users are unable to place the mouse adjacent to the keyboard because of space constraints or work area layout or barriers.

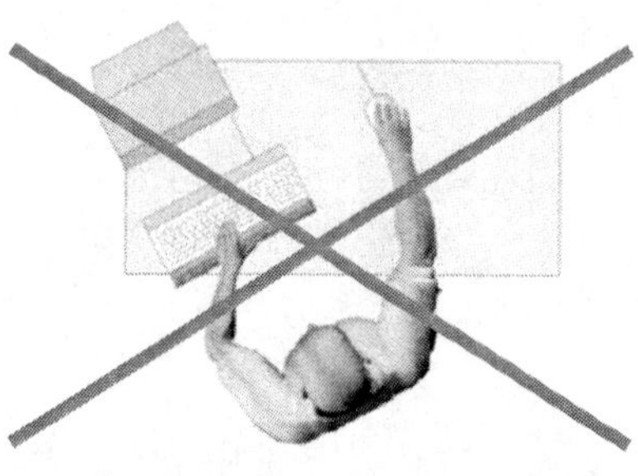

A mouse location away from the body causing an awkward arm posture.

Mouse placement is far more important than many people realize, especially for software applications that require frequent mouse movements and clicks. Though the postural deviations, movements and forces required to operate a pointing device may seem insignificant at first glance, these factors, compounded over hours, days, weeks and years, become a significant risk to comfortable and safe performance for many computer operators.

5.2.4.5 Keyboard and Pointing Device Height Considerations

The keyboard and pointing device should be on the same level, ideally right next to each other, at a height that minimizes wrist angles (flexion and extension). In general, if the arms are unsupported, the lower arm should be parallel, or nearly so, to the floor, and the upper arm should be vertical, or nearly so. The root goal is to minimize unsupported reaching and awkward wrist, elbow and shoulder angles.

5.2.4.6 Placement of Other Work Tools

Depending on their tasks, computer operators may need additional equipment to perform their job. A telephone, for example, is typically a required tool for office workers. An employee performing frequent transcription from paper copies can benefit from a document holder. Someone doing frequent calculations might need a separate 10-key or desktop calculator. Each of these tools will occupy work space, and their placement is directly related to the frequency and duration of their use.

5.2.4.7 Telephone

For occasional telephone use, the phone can be located at full arm's length so that more frequently accessed tools are in the primary reach zone. For frequent telephone use, the phone should be closer, in the primary reach zone. There is no definition of occasional or frequent, so these distinctions are left up to those most familiar with the work.

Cradling a traditional phone between the head and neck is likely to lead to problems. Therefore, for frequent phone users in particular, a headset is a smart investment.

Avoid cradling the phone between the head and neck.

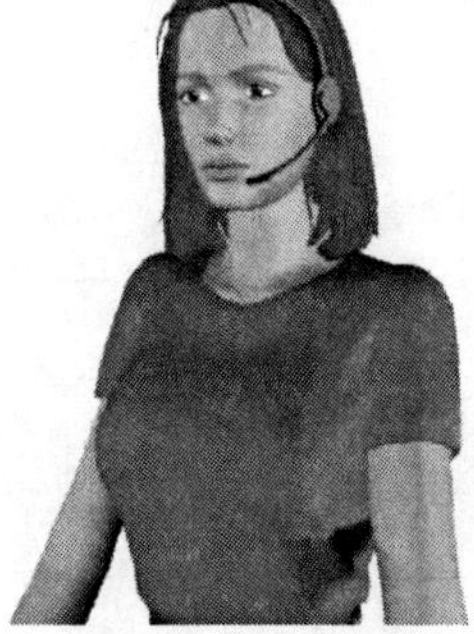

A headset is a smart investment for frequent telephone users.

5.2.5 Static Posture and Duration Are Often the Culprit

Most computer work requires low force exertions in postures that would normally be of little concern. After all, when compared to work in manufacturing settings, for example, how could a computer operator complain, let alone get injured? The answer lies in the duration of the postures and exertions. It's easy to turn one's head to the side. However, when an operator works each day, day after day, week after week, month after month in that slightly deviated posture, comfort, productivity and quality of work product are likely to be affected. The same principles apply to the wrist, arm, shoulder, back and legs.

Introducing more movements, either through task or equipment design, inherently reduces the duration of any given static posture. Without training, even the best furniture and equipment investments will never reach its desired ROI.

5.2.6 Adjustability Is Important

Without adjustability, almost no operator will fit a workstation well and be able to perform at his or her best. Therefore, furniture and work tools must be adjustable, and operators must know how and understand why adjustments should be made.

5.2.7 Furniture Dealers May Not Have Your Best Ergonomics Interest at Heart

Many furniture dealers purport to be experts in ergonomics or profess that their solutions are "ergonomically designed". Not surprisingly, these claims may not be true. As always, *caveat emptor* – "buyer beware." The following article exemplifies the conflict many furniture dealers have when it comes to making recommendations.

Reprinted from Ergonomics Today™. Used with permission.

Are You Getting an Ergonomic Assessment or a Sales Pitch?

July 25, 2001

I was recently asked by a client to evaluate an evaluation they had received. Yes, that's correct, evaluate an evaluation. This person had received an office workstation evaluation by someone claiming to be competent in ergonomics. The problem was, even to the office worker, the final recommendations seemed a little odd and extremely expensive.

The office workstation had some typical problems. The keyboard and mouse were not at the same level, the keyboard and monitor were not in line, and there were some awkward reaching postures. Much of this could be fixed by measuring out the workspace and rearranging things including file drawers and reference material. The worker stated that it was financially possible to rebuild some of the working areas, and we discussed economical and practical solutions that would incorporate this reconstruction.

The client then asked if I would take a look at the previous evaluation recommendations. Always optimistic that those claiming to be ergonomists actually have a high level of knowledge and competence, I eagerly reviewed the evaluation. I was sadly disappointed.

To begin with, it was nearly impossible to find the actual "evaluation". I first had to wade through pages and pages of laminate counter top selection and matching chair fabrics. I finally found what looked to be a page including some measurements. "Are these the measurements of your workstation?" I asked. "No," replied the worker. "Those are the dimensions of each desk section and the column on the right is the price of each section."

Trying to quickly regain my composure after having seen the numbers in the right side column, I asked where the rest of the evaluation was. With a disappointed and somewhat confused look, the client replied, "That is the evaluation." And with that, the full extent to which this client was deceived and misinformed becomes apparent.

The "evaluation" this person had previously received was "free", a complimentary service performed by a representative of a very large furniture maker. I do not know what qualifications the evaluator claimed to have, but from his obvious neglect of some very fundamental ergonomics concepts, I do not think they spanned even reading a textbook on the subject. This is exceptionally distasteful as he represented himself to the public as an "ergonomics expert" and not a salesperson. Selling expensive, name brand and possibly unnecessary products, it seems, was the sole purpose of the so-called evaluation.

So is there a moral to this tale? Yes, and more than just the questionable business practices of the furniture maker here! Qualified ergonomists need to make every effort to uphold the integrity of the profession. Many times that will include helping clients and the public understand the difference between what is an ergonomics evaluation and what is only a sales pitch. This does not mean that new or different products are not an appropriate solution in some cases.

In any evaluation, it is important that the consumer ask questions. If products are recommended, ask exactly what problem each product addresses and if there are price ranges available. Ask if there is a trial period allowed for a product. And as with any professional service, don't be afraid to ask about the credentials of the person performing the evaluation.

- Rachel Michael, M.Sc., AEP

This is not to say that all furniture vendors are deceitful. In fact, some specialize in ergonomically designed equipment, and have the necessary training and background to properly advise a customer on the match between people, equipment and task. Find out whom your local ergonomists prefer to deal with, and seek their advice.

5.2.8 It Doesn't Have to Be Expensive

Practical solutions are often the best. Many times it's as simple as rearranging workflow, equipment and tool locations, or making a few adjustments. Sometimes it's a matter of training. Sometimes it's enough that "someone cares".

5.2.9 The Task Defines the Work Tools

An office worker that only occasionally spends time on the phone does not necessarily need a headset. Someone who does not transcribe from text documents does not need a document holder. Computer operators that are able to adjust their workstation to a comfortable and productive set-up while still resting their feet comfortably on the floor do not need a footrest. In fact, providing unnecessary tools and equipment only causes clutter and interferes with the preferred arrangements of the necessary components.

All equipment and work tools should be selected based on the task demands.

5.2.10 Beware the Label "Ergonomically Designed"

A product that claims to be "ergonomically designed" may or may not be so. There is no recognized standard to test whether something is ergonomically designed, so any manufacturer or sales representative can make this claim, true or not.

In fact, some designers and manufacturers do go to great lengths to design for specific populations, tasks and environments. But many don't. The ultimate fit test is: Does the product reduce specific risk factors or produce specific increases in human performance when being used for a specific task? If yes, it may be ergonomically designed, as long as it is used by people in its target population in the task and environment for which it was designed.

Reprinted from The Ergonomics Report™, Vol. 2, No. 10, October 2003. Used with permission.

Beware the Label "Ergonomic"

You see them everywhere: ergonomic pens, ergonomic scissors, even ergonomic bras. But how ergonomic are they really?

A truly "ergonomic" tool will consider more factors than just padding on a handle. User anthropometry, work space design, task and usage as well as human capabilities will be taken into account to design something that works for the user.

This isn't to say that all tools labeled "ergonomic" are bunk. If a tool is used for the purpose and environment intended, and by a person who fits the anthropometric profile, chances are the tool might actually be ergonomic. But that still doesn't make the tool ergonomic for everyone or for every situation.

Currently there are no standards that a product must meet before it can be labeled "ergonomic." So, while some products have gone through human-centered design and user testing, others may be labeled "ergonomic" only for marketing reasons. In the absence of any such standards, ergonomics becomes a "buyer beware" market.

Rule of thumb: if the tool results in a reduction of risk factors like forceful exertions, awkward postures, repetitive exertions and exposure to environmental factors like extreme heat, cold, humidity or vibration while also improving work performance, it's probably ergonomically designed. But a tool that may work for one person may not do the trick for another. The tool needs to fit the worker, the task and the work environment.

Applying the tool to the environment, for example, might mean choosing a pistol grip rather than an in-line grip tool for posture considerations, or vice versa. Ultimately, if a worker has to work at some sort of angle to his or her task, it's the tool that should do the bending, not the worker. Other issues to consider are grip sizes; if the grip is too big or too small for the user's hand, it can impact the stability of the tool. And tools that aren't properly balanced may require additional effort to control, and thus impact other risk factors as well.

Ergonomic or not, tools must be regularly maintained. A custom-designed pair of scissors built specifically for the task and the user do little good if they're dull and require increased effort to operate.

Hopefully in the near future, an "ergonomic" label on a tool, be it a backpack, an egg beater or a drill, will actually mean something, more like the term "organic" - a regulated term that can only be used if the product fits certain pre-set criteria to earn that label. But until then, it's up to the customer to make sure the tool is living up to his or her expectations and capabilities by reducing potential risk factors while still serving a productivity-related purpose.

- Jeanie Croasmun

5.2.11 Don't Forget About Software Usability

Software interface design can be one of the greatest barriers to worker productivity and work quality, and cannot be ignored when evaluating or designing computer work. For example, font types, sizes and color schemes can have a dramatic effect on visual strain and appropriate monitor placement; and software navigation and interface designs can dramatically affect training requirements and costs, as well as error rates.

Human Computer Interaction (HCI) and software usability is an important area of study and must be considered in any comprehensive approach to office ergonomics, but is beyond the scope of this book.

5.2.12 Don't Forget About Lighting and Glare

In order to effectively accomplish a task, a worker must be able to adequately view the work. This means lighting in the office must be sufficient. However, too much or inappropriately directed light can cause glare. Glare occurs when there is a large difference in the amount of light, or luminance, of an object and its background. Glare can come from bright sunlight, lights in the field of view or be reflected from a shiny surface. In addition to causing discomfort and eyestrain, glare can force an employee to assume an awkward posture to view the monitor around the glare.

Reducing glare may be a matter of moving a monitor or changing its angle relative to light sources, such as a window. It may also be possible to reduce glare caused by overhead lights by adjusting the tilt of the monitor or using light fixture baffles to deflect the light at its source. Screens that fit over the monitor to block out glare are also available.

Task or desk lighting, in addition to overhead lighting, gives employees the opportunity to control some luminance around the work area. A desk lamp that is adjustable in position and brightness may help improve productivity, quality and reduce some MSD risk factors.

A variety of literature is available on appropriate lighting levels for different work areas. If designing a work area, it is suggested that specific references on the subject be consulted.

Reprinted from Ergonomics Today™. Used with permission.

Stiff Neck, Headaches, Productivity Down? Check the Lights

May 5, 2003

Think the only problem with poor lighting is that it makes you sleepy? Not so, says the Canadian Center for Occupational Health and Safety (CCOSH). Poor lighting can also lead to neck and shoulder pain, eye fatigue, headaches and even poor productivity.

According to the agency, workers receive over 85 percent of their information through their eyes, and when the lighting conditions are poor, that information runs a higher risk of being processed inaccurately.

Common poor-lighting complaints from workers include eyestrain, eye irritation, blurred vision, dry and burning eyes, and headaches, but other issues including posture and overall safety can be impacted by poor lighting conditions as well. Poor lighting can mean anything from lighting that causes glare, to overly bright lights, improperly placed lights or dim lighting.

Posture becomes an issue with lighting when workers assume awkward positions in order to more clearly see the work, and safety issues arise when workers have problems visually determining the position of a tool or the shape or speed of machinery.

CCOSH offers some suggestions for improving lighting conditions in the workplace including using filters to diffuse overhead lighting, covering windows with adjustable blinds, and painting walls, floors and furniture with a matte finish.

Source: CCOSH Health and Safety Report, March 2003 (online)

– Jeanie Croasmun

5.2.13 Ergonomics Requires a Systems Approach

A complete approach to office ergonomics requires an understanding and consideration of the physical and psychological attributes of the person or population of people who will perform the jobs; the design and arrangement of the workstation and furniture; computer hardware; computer software; workstation accessories and tools; the tasks required to perform the different job; the work environment, including noise, temperature, ventilation and lighting; as well as management and organizational methods and constraints.

The interaction between these general topics is critical and will define the postures, forces and repetitions assumed by the person(s). Further, since a modification in one area may have significant effects in another, no single change can be performed without considering the overall effect. For example, lowering seat height so that someone may comfortably rest his/her feet on the floor may force a stressful upper body posture if the monitor position and table or keyboard height are not adjusted in concert (this is often a good reason to provide a footrest instead, or provide an adjustable work surface).

5.3 Fitting an Individual

Though still highly desirable, furniture and equipment that can be easily adjusted by the user is less important when fitting an individual than it is when fitting a population of workers. Especially when budget conscious, an individual workspace that will not be used by others can be effectively set up to approximate the preferred working arrangements that match the work tasks, equipment, work tools and operator preferences.

5.4 Fitting a Population

It is essentially impossible to fit the working population without providing furniture and equipment that is easily adjustable by each user. Office environments with shift work, shared workstations, heavy turnover or worker rotation are example situations were multiple adjustment options are required to provide fit for varying people.

5.5 Ergonomics and an Aging Workforce

Reprinted from The Ergonomics Report™, Vol. 2, No. 11, November 2003. Used with permission.

Ergonomics and An Aging Workforce

It's inevitable: keep a car long enough and it's bound to have problems. The once-vigorous machine becomes a little sluggish on that long uphill or it doesn't handle curves with quite the verve it used to. For most of us, when a car starts to show signs of age, it's time to trade it in for a new model.

The solution to an aging workforce, however, isn't that simple.

It's estimated that in the next five years, 25.2 million members of the U.S. labor force will exceed the age of 55 and will remain active in the labor pool; that's over 16 percent of the workforce. By the year 2025, that number rises to over 20 percent and the percentage of workers in their 20s, 30s and 40s decreases. Wisdom may accompany age, but so do physical limitations. Just like an older car, the body isn't the smoothly functioning machine it once was.

The problem arises when the workplace, built to accommodate the 25 year old engineer who designed it 30 years ago, is still being used by the same, now 55 year old engineer today. The suit he wore when he designed the place probably won't still fit him the way it once did; unsurprisingly, neither does the workplace.

But does that mean it's time to trade in the old workforce for a fresh batch of college grads? Of course not. It means it's time to take a look at the work and the workplace itself to determine how to tailor it to the aging worker as well as the younger worker, increasing productivity and efficiency and reducing injury potential across the board. In essence, the workplace grows up with the workforce, and with ergonomics interventions, accommodating workers, young and old, means the business has a chance to mature, too.

The Graying of the Workforce

The media might call 40 the new 30, but any age beyond 65 seems to be the new retirement age. Workers are retiring later. With advances in health care and life expectancies being higher than anyone would have dreamed a few years ago, older workers are rapidly becoming a fact and a necessity.

Workers are retiring later and staying productive longer. A 2002 poll by the AARP of American workers between the ages of 45 to 74 found that less than a third are planning to stop working when they reach their retirement years. That means more than 66 percent of today's workers are in it for the long haul and aren't planning to go anywhere – 84 percent of the respondents even indicated that they'd continue to work if they won the lottery.

But while today's and tomorrow's older workers have the drive necessary to remain productive, what they may not have is the physical ability to achieve their full potential unless workspaces change to accommodate them. Physiologically, older workers are faced with a variety of time-induced limitations including failing eyesight, hearing, flexibility and dexterity, slower response times, back problems, and a greater potential for injury. It's estimated that the body starts to age at about 30, that the eyes will go between 40 and 50, yet workers may still plan to contribute effectively for another 20 years or more. With all of these facts of old age stacked against the workforce, how is it possible for an employer to keep up and remain profitable? By looking at each unique aspect of aging, one at a time.

Answers In Sight

Have a pair of glasses for reading, for seeing far away and maybe one more for in-between? Then you're probably over 40.

Just when it's time to settle down and focus on what's ahead, our eyes start to play tricks on us, blurring those minute details that were once so painfully clear. It's a fact of life: by the time a worker reaches age 50, there is almost a 100 percent chance that he or she will have glasses and will have some sort of struggle with the eyes. And usually, it's what's up close, like a book or a computer screen or a small detailed part on an assembly line, that suffers the most.

"We're working with the same bodies, but our focusing ability decreases. Our maximum focusing ability is at age 10. Then the lens thickens and it becomes harder to focus as we get older," says Dr. Jeffrey Anshel, a California-based doctor of optometry and the author of a number of books on vision and ergonomics. Ultimately, the vision suffers and depending on the work environment, even with corrective lenses, that can be a problem for the worker.

For Anshel, whose books cover everything from general eye health to computer vision syndrome, the aging eye and its many potential problems including nearsightedness, farsightedness and even cataracts, can be particularly tricky to treat. And when the worker is in an office environment, a simple bifocal, trifocal or progressive lens may not be enough to compensate for the eye's changes.

Anshel notes that vision tests typically measure for distance at 20 feet and near vision at 16 inches. "Doctors will prescribe a progressive lens," says Anshel, a type of lens that has corrections for close viewing, far viewing and intermediate viewing situations which is where the computer normally lies. But, says Anshel, that configuration doesn't work well for the computer user since the intermediate viewing lens is normally the most narrow part of the progressive lens and today's computer screens are anything but narrow.

Ultimately, that means that an older worker who spends any time on a computer could need more than just a one-pair-does-all set of glasses. Struggling to see letters clearly or battling vision-related headaches doesn't improve the productivity or well-being of any worker. For an older worker to see the monitor as clearly as possible, it might mean a separate pair of glasses that concentrate on the computer's mid-range as well.

Additionally, Anshel suggests looking at lighting in an office. "Most offices were designed for paper. But [with computers] we've gone from a horizontal work surface to a vertical one. It's a challenge, however; people are still using paper." But to eliminate glare, workspace lighting should be appropriate for the task; if the task sits on more than one plane, like computer work and paper work, workspace lighting needs to adjust accordingly.

While computers and aging workers' eyes may need a little tweaking to work together effectively, an older worker with a computer can be just as effective as the younger worker. "Computers have allowed us to be more productive. This is the first generation to take a 100 years to turn 50. We're not our parents. [Workers today] will be functional longer," says Anshel.

Designing for Everyone

At birth, males outnumber females. Between the ages of 60 and 64, there are only 89 males left for every 100 women and by the time they reach 80 or above, only 53 men remain for every 100 women. Men simply die sooner.

For the workforce, that means designing for age is slightly skewed towards designing for women. Smaller size, less bone density, less strength and more prone to osteoporosis are all factors that can impact workspace heights. Additionally, women report more incidents of musculoskeletal disorders (MSDs), although the reason for this is still unclear.

Teaching and nursing, already predominantly female professions, may benefit the most from ergonomics interventions intended to work with aging female workers. Both professions are already seeing shortages of employees, and retaining workers past traditional retirement ages may be one way to keep these sectors of the workforce from even greater employee shortages. Currently, 21 percent of teachers in the U.S. are between the ages of 50 and 59.

As any worker ages, male or female, height can decrease by as much as three inches. Dexterity becomes limited, flexibility is compromised and response time becomes greater.

So creating an ergonomically sound workplace for older workers isn't just a matter of designing the workplace for older women, it's a matter of making the workplace more agreeable for everyone in the workforce. "The goal should be to make jobs safe for everyone," Dr. Laura Punett, senior associate of The Center for Women and Work at the University of Massachusetts-Lowell said in an April, 2003 interview with *The Ergonomics Report*. "Some key [design] issues include designing the physical dimensions of workstations, equipment and tools, and personal protective equipment to fit the range of body sizes and shapes found in our increasingly diverse workforce," Punett said.

Making changes that work for everyone means injuries can be reduced, productivity can be increased and the job can fit workers of all abilities. After all, an older worker that knows the job and can offer the workplace the benefit of experience is a valuable resource. This means that even though a worker is not the newest model, he or she can still contribute just as well, if not better, than a worker fresh off the line.

- Jeanie Croasmun

5.6 Special Considerations

5.6.1 Shift Work

By its very nature, shift work requires different people to occupy the same workstation. As discussed previously in Section 5.4, a high degree of adjustability, and the training to understand and use those adjustments, is required if each worker is to be able to achieve comfortable and productive working postures for their given tasks.

5.6.2 Accommodations

The Americans With Disabilities Act (applicable in the USA), discussed in detail in Section 4.3.3, encourages employers to provide accommodations for employees with physical and mental disabilities. In such cases, it is common to provide specialized equipment and physical workstation designs that fit a particular individual with a particular task in a single workstation.

5.6.3 HIPAA

The Health Insurance Portability and Accountability Act of 1996 (HIPAA, applicable in the USA) is designed to protect the medical records and associated personal information of individuals. The act places certain privacy restrictions upon those organizations that handle personal health information. Providing ergonomics services can, at times, include gleaning personal medical information from individuals. In order to protect yourself, your company and your client, familiarize yourself with the information contained in HIPAA prior to providing service.

For organizations that handle personal medical information as a normal part of their operations, HIPPA requires that steps be taken to ensure that the information is kept private, even between workers in the same organization. (e.g., the cubicle next door). This implies that adjacent workers should not be able to overhear each other when verbalizing personal medical information, and passing workers or visitors should not be able to see documents or computer screens containing protected information. These same protections apply to environments were other patients or visitors may overhear or see another's medical information.

The provisions of HIPPA may therefore result in office designs, layouts or equipment selection that present special challenges for ergonomics. For example, a privacy screen on a computer monitor that protects information from passersby may also change the way the computer operator sees the screen.

5.6.4 Influencing Purchasing Decisions

Furniture and computer equipment are capital purchases and typically require documented cost-benefit analysis in order to justify the expenditure. If, for example, a purchasing decision maker is able to purchase basic computers for $800, an IT department must be able to specify performance and hardware requirements, complete with documented reasons to upgrade from the $800 basic model, in order to justify a more expensive purchase. Likewise, if you recommend specific equipment that is more expensive than a cheaper, basic model, then you must be prepared to specify your reasoning and your expected payback for the added cost.

Projecting savings in terms of injury reduction can be difficult, although the rising costs of risk management and human resources (e.g., Workers Compensation insurance, direct and indirect costs of injuries, lost work time, high employee turnover rates, new hire training, etc.) are reaching the ears of corporate decision makers. A well-planned ergonomics-driven purchase should, however, be able to demonstrate paybacks in traditional purchasing decision points of productivity, quality, employee retention, reduced turnover and improved morale. These types of business metrics may prove to be more successful influencers than health and safety alone.

5.7 Ergonomic Furniture and Equipment

5.7.1 Introduction

As mentioned throughout this book, simply labeling a product 'ergonomically designed' does not make it so. Office furniture and equipment, in recent years, has seen a boom in products claiming to ergonomically ease pain and increase productivity. The key to successfully acquiring the appropriate furniture lies in critically defining the task and the population of users. Take, for example, a document holder. Should all workstations in a facility be equipped with a document holder to be considered 'ergonomic'? If so, what type of document holder should it be, one that attaches to the side of the monitor or one that sits in front of the screen? The answer to this is, "it depends". It depends upon the tasks that must be performed and the people who will be performing them.

If a worker only retrieves information from a computer for customers who call in on the phone, and this task requires no paperwork, a document holder may not be necessary. On the other hand, if an employee is required to input information from a printed page into a computer a document holder may be a very appropriate tool that could improve productivity, reduce errors and reduce MSD risk factors. In the later case, a document holder would likely be a good engineering control.

A common mistake made when purchasing furnishings or equipment for an office is choosing only one type of product. For example, a purchaser may see that a certain trackball is less expensive than a mouse and order enough for all users, or a chair with a high back may be ordered for all workstations. In addition to personal preferences that affect comfort and productivity, these products may not be appropriate for the tasks performed or people performing them. One solution to this is to have a product display area where a few styles of products, such as a mouse and trackball; high back and low back chair; or different styles of headsets, are available. This allows employees to try a product before committing to it, which can immediately save capital expenditure and lead to increased productivity through a good match between the user and the equipment.

A short description of some common office furniture and equipment is provided in the following sections. This is not a recommendation of specific products or manufacturers. It is strongly suggested that you understand and communicate your specific needs to vendors before making a purchase.

5.7.2 Chairs

Many adjustable chairs are available today. In addition to an adjustable chair, a person may need a sized chair. Some manufacturers are making chairs in sizes like small, medium, large and XL. To ensure a good fit, a variety of chairs to choose from is recommended.

An office chair should have some basic features. Some favorable chair features include:

- A variety of adjustments to help fit the worker and the task.
- Seat height and angle adjustments that allow comfortable access to the work area.
- Seat pan depth adjustability that allows a worker to reach the backrest without creating pressure or discomfort behind the knee.
- A backrest adjustable in height and angle to fit the curvature of the lower back.
- Comfortable for each user.

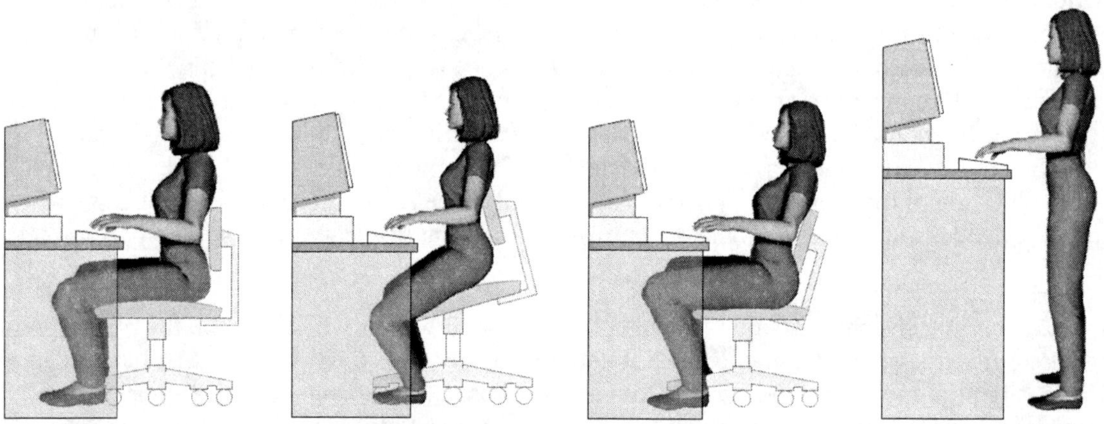

Sitting upright | ***Sitting forward (decline)*** | ***Sitting back (recline)*** | ***Standing***

Additional adjustments might include seat tension, lumbar support firmness and armrest height if the chair has armrests.

If a worker cannot rest his/her feet on the floor while seated at a comfortable height, a footrest may be appropriate.

Reprinted from Ergonomics Today™. Used with permission.

Is an "Ergonomic Chair" Enough?

July 10, 2001

A golden rule in ergonomics is "one size does not fit all," and adjustability is a key feature that can transform a typical chair into an "ergonomic chair." But, as many companies have learned the hard way, an adjustable chair is not always enough.

Successful ergonomics requires a systems approach. An office chair is one component in the office system, and the person that sits in it is another. The stapler, telephone, computer, mouse, keyboard, desk and other work tools are additional components in the system. How the person interacts with those components as he or she performs required work tasks is yet another part of the overall system. Selecting a chair without considering the rest of the system components and work tasks usually leads to a misfit and may introduce new, unforeseen problems into a workstation.

In order to fit the broad working population, a computer workstation must have multiple component adjustments, not just chair adjustments. If the chair is the only easily adjustable feature, most people will still not be able to achieve their most efficient, productive, safe and comfortable working postures. This problem is not always obvious to the casual observer, and it therefore persists in many companies.

Anthropometry, the measurement and study of human body dimensions like height, weight, reach lengths and eye heights, is one of the most important data sets applied by ergonomists working to fit people with their physical environment. It's easy to see that people vary in height and weight, but it's less obvious that we also vary in many other dimensions, such as leg and trunk lengths. For instance, some races have proportionally longer legs than other races, and even within the same race, specific body dimensions vary substantially from person to person.

To illustrate, imagine two women of the same height, standing side-by-side. We measure the hip height, which corresponds to leg length, and find that one has legs three inches longer than the other. Next, we measure seated knee height (popliteal height, as the dimension is technically called), and we find yet another difference between them. Finally, we ask each woman to sit in a chair and adjust the chair height so that each may comfortably rest her feet flat on the floor (shoe height introduces yet another variation). Even though they are the same standing height, we'll find that the seated height may differ substantially, because their leg and their torso lengths are different. Important dimensions like seated eye and elbow heights will also be different, which means each will have different height requirements for their keyboards, mice and monitors.

Contrary to claims otherwise, there is no perfect way to sit, nor is there a perfect way to arrange all workstations, even if the same job is being performed at each. Natural variability in people dictates that the system must be flexible in order to meet the unique needs and dimensions of all the different people that may interact with the system. An adjustable chair provides some of that flexibility, but not enough to accommodate the full, multifactorial spectrum of anthropometric variations. Simply put, the location of the computer keyboard and mouse need adjusting independent of the chair, as may the monitor.

The extent to which a system should be adjustable depends upon the nature and diversity of tasks to be performed at the workstation, and whether one individual person or a population of workers is being fitted to the system. If only one person will use the workstation, a one-time set-up and custom adjustment may suffice. If the workstation will be used by different people (e.g., shift work or a shared nursing station), a system that can be easily adjusted by each worker is recommended. Even without shift work, employee promotions, transfers and turnover will result in different people using the same system, and adjustment will still be needed but at a lower frequency.

There are several common and relatively inexpensive workstation components designed to accommodate variations in size and preference: footrests, keyboard and mouse platforms, and monitor arms or risers. When work surface or monitor heights are "too high," a person may benefit from a footrest, allowing her to raise her seat, yet still be able to comfortably rest her feet. When desk height, and therefore the keyboard and mouse that rest on it, is "too high" or "too low," an adjustable keyboard and mouse platform may help. And when the monitor height is "too high" or "too low," monitor adjustments may help. Each of these accommodations have strengths and weaknesses.

- **Footrests:** Relatively inexpensive and easy to use. Drawbacks include the potential for clutter under a desk, or even a tripping hazard, and may restrict freedom to move and achieve desirable dynamic postures throughout a workday. On the other hand, a footrest is a simple way to accommodate smaller people in particular, who often have difficulty keeping comfortable leg postures. If a footrest is selected, adjustable height and angle are recommended.
- **Keyboard and mouse trays:** These are available in a variety of shapes and sizes, and with a variety of adjustments. To be effective, the tray must be large enough to accommodate both the keyboard (and palm rest if desired) and the mouse, side-by-side, and on the same level. Easy height adjustment and positive and negative tilt are recommended so that the user can reduce wrist deviations in a comfortable arm posture. There are drawbacks with such platforms. For example, they usually extend out from under the desk surface, pushing the worker further away from the work surface, and they can present an obstruction under the desk, sometimes hindering leg access and movement and interfering with chair arms. Quality is important when selecting these devices; be sure that the person can type and operate the mouse with stability (i.e., avoid "flimsy" models).
- **Monitor arms and risers:** Monitor heights can be raised cheaply by using such things as old phone books. If a more tasteful solution is desired, there are a variety of monitor raising products on the market, including stackable trays or adjustable arms. Lowering a monitor is limited by the work surface height and monitor design, unless a recessed monitor desk is selected. The advent of the flat panel display makes monitor placement and adjustment easier.

Ergonomists often feel that these types of accommodations are at best a retrofit to a system with deeper problems. Ultimately, the best solution is to provide an easily adjustable work surface, not just adjustable add-ons. In fact, work surfaces that adjust through a height range from seated work to standing work are no longer uncommon. Such systems allow workers to adjust to different heights for different tasks and different people, and to make postural adjustments for comfort and fatigue over the course of the workday.

In many cases, cost is the deciding factor when selecting between add-ons or fully adjustable work surfaces. However, more manufacturers are offering fully adjustable systems or work surface adjustment retrofits, and competition is driving prices down. Furthermore, a larger upfront investment can result in significant future savings.

"Work tools" is a general term that describes the accessories that support a worker's duties. A telephone, stapler, document holder or even a writing pen are all examples of work tools in an office environment. In the same sense that a carpenter needs proper tools to perform quality woodwork in a productive manner, a call center worker needs the right tools for his or her job. Proper tool selection can be the difference between success and failure, both for the company and the individual.

- Peter Budnick, Ph.D., CPE

5.7.3 Computers and Accessories

5.7.3.1 CPUs

The CPU or processing part of the computer should be located where it does not obstruct leg or arm room. The following article discusses some of the issues concerning placing a CPU under the desk.

Reprinted from The Ergonomics Report™, Vol. 2, No. 10, October 2003. Used with permission.

Under-the-Desk CPUs

Question

Does anyone have any information regarding the ergonomics associated with IT technicians servicing computers under desks? Computer users always seem to want their CPUs placed under the desks but then the techs who do maintenance on the CPUs have to go under the desks to service the computers.

Reader Opinions

Reader opinions are extracted from a detailed thread on ergoweb-L, *Ergoweb's on-line ergonomics discussion list (http://www.ergoweb.com/resources/discussions). Visit and join the discussion to read, learn, view and get ideas and answers to tough questions regarding ergonomics in the workplace.*

There is one solution that may fix your problem. You can implement remote workstations. You can add a transmitter adapter card in the PC that connects to a small receiver box at the desk that provides access to audio, mouse, keyboard and video ports the user needs. The adapter connects to the receiver box via CAT5 cable and can be up to 100-200 yards/meters away. This provides good security, much easier serviceability for techs, reduces space required at the desk, but takes the system away from the user which can be perceived as losing some control. Amulet-Hotkey and Wey Technologies are companies I am familiar with that have this technology. Let me know if you would like more info.

- Dan Kelaher
NCSU Ergonomics Lab
IBM Human Factors

In most cases where users say they want CPUs placed UNDER their desks, I believe they really mean they DON'T want them ON the desk, for obvious reasons.

PCs have been an integral part of office work for almost a quarter of a century (well, perhaps two decades anyway), yet we still see the vast majority of "office" furniture made without proper facility for CPUs, monitors, cable management (other than a hole in the desktop), etc. The most common, and often only, token tip-of-the-hat to screen-based office work is the provision of a keyboard tray (which more often than not is unnecessary anyway). Often when provision is made for the CPU, it is in the form of a narrow, vertically-oriented slot with almost no accessibility to the rear of the CPU without crawling on the floor under the desk.

CPUs and their multiple cables have been with us for decades and will likely be with us for quite some time yet (the laptop and wireless revolutions notwithstanding) ... When will furniture designers and manufacturers start to think about and produce designs that can accommodate CPUs in a way that makes them:

- easy to emplace and install;
- easy to reach for the user (for access to disc and CD drives and controls, see indicator lights, etc.);
- easy to attach or troubleshoot the multiple cable connections;
- easy for technicians (or us) to check, maintain, etc.

... and do all that without either demanding awkward calisthenics or encroaching on the workspace (both under and above the desktop) necessary for safe and effective human performance?

Of course, computer manufacturers could also shoulder some blame for continuing to put all the "messy stuff" (cables, connections) out of sight at the back when, more importantly to us, to the user and to the techs, they are putting it out of REACH as well.

I know there are some alternatives, but in my experience, they are few and far between. Where are the creative designs and the better solutions??

Perhaps we (ergonomists) should also shoulder blame for not speaking loudly enough and in the right direction ... any suggestions there?

– Max Hely
Safety Science Associates Pty., Ltd.
Sydney, Australia

Since workstations have shrunk in recent years, locating the CPU on the desktop has become increasingly unpopular.

I have recommended hanging CPUs from desktops, rather than placing CPUs on the floor, to avoid having workers bending over to start their machines and use digital media. This recommendation requires finding out the weight specification of the desktop to ensure this option would not exceed it, and ensuring that the user still has adequate turning radius to swivel at the desk.

Historically, the workplace facilities groups and their furniture consultants have performed quite non-analytically regarding the layout of computer systems; I suspect we all remember the typical layout of the rectangular CPU being plunked under the monitor and calling it good, requiring users to flex or extend their necks to see the information on the monitor with painful results.

For the amount of use and overwhelmingly stimulating effects the computer gives to workers, it would be quite supportive of the end users if we could get buy-in on optimal equipment layouts to support productive output. In my experience this means contracting with facilities groups for ergonomic standards PRIOR to the new equipment purchase and training upper level management about ergonomics.

- Beverly Burke, RN, CMA, CIE

One argument we have used to get the machines off the floors is "cable management." An accident involving an employee tripping as a result of getting entangled in the wire between her mouse and CPU (our tech support does not include many wireless options) reinforced this argument.

There are other concerns for the CPU such as the lack of venting under a workstation, the accumulation and subsequent degradation of the CPU due to dust or efforts to deal with dust like carpet cleaning, and inability of the cleaners to vacuum due to the accumulation of things under the work surfaces (including personal belongings). These points apply equally to anyone working on the CPU.

The argument to have CPUs under the desk has been mitigated largely by the advent of the flat screen monitors now being less expensive to buy. In fact, from an ergonomic point of view, our workplace health and safety committee forwarded a recommendation to the policy committee that, in future, only flat screen monitors be purchased. While the recommendation was based on the value to the employer for fewer eye strain problems, an added argument is the additional space available on work surfaces, the lack of which is one of the arguments for putting CPUs on the floor in the first place.

- Canadian worker

Maybe one should look at the mid to upper end home office market for under desk solutions. My experience with home office furniture is that the client will not tolerate the wiring "junk" that comes with the CPU. They also demand accessibility to the CPU. Consequently, several designs have been available with generous accommodation for under desk CPUs as well as providing full extension platforms that provide effortless access to the rear of the unit. Enclosures are also part of the game because "out of sight" is another criteria that influences home office furniture design. The reason well designed furniture is usually far better looking than most computer equipment.

- Jack Kelley, principal designer
Studio 222

"Under the desk" does not have to mean "on the floor." Weber Knapp sells mounting hardware for CPUs that suspends them underneath a work surface and allows them to be slid out on a track for improved access to cables, etc. They are listed under "Ergonomic Products–Accessories" on their website; other vendors may have similar products.

A tech will still have to remove the CPU for installing cards or some types of maintenance, but this is a small percentage of the time compared to the worker being seated at the workstation. Being off the floor improves ventilation and allows for vacuuming.

– Philip Jacobs, MS, CPE, CSP
Jacobs Consulting, Ltd.

Ergoweb Answers

Computers are tools, at least for the computer user. But for the person who maintains the computer, trying to reach the under-the-desk computer is a task that needs fine tuning.

There are any of a number of reasons why computer operators would want their CPUs under the desk, but usually the CPUs are stashed out of the way to free up valuable desk space and eliminate desk clutter. However, just because an option looks good (hiding the CPU under the desk, for example) doesn't mean it's a good ergonomic option.

Under-desk CPUs can cause problems for desk workers in the form of loss of leg room or awkward postures from bending down or leaning over to use CD, floppy disc or DVD drives. And IT techs whose jobs find them accessing those under-the-desk, ground-level computers each day can also face risk factors associated with the awkward positions they have to assume to reach the computer. If the risk of injury weren't enough, there's a productivity side to relocating computers to easy-to-access locations for the technician: imagine how much more quickly an IT tech could complete the task if the CPU were readily accessible.

The solution might likened to an auto mechanic's shop. While a mechanic could bend and twist to scramble under a car, lifts are used instead to position a car at a level that's more comfortable and agreeable to the mechanic. Apply that same theory to an office environment and, rather than an under-the-desk computer that requires the maintenance tech to come to it, the computer itself is positioned in a more comfortable location for maintenance with the help of another tool, like some of the ones recommended by our readers (above).

– Jeanie Croasmun

5.7.3.2 Input Devices (Mouse, Trackball, Touchpad, etc.)

An input device such as a mouse, trackball or touchpad may also be part of the workstation. These items should be placed right next to the keyboard and as close to the body as possible.

A mouse placement that requires reaching

A mouse placement that doesn't require reaching

There are many styles of mice, track balls and touchpads. Some are activated with wrist movement, some with arm movement and some with finger movement. Some have a 'clicking' function that is operated by the fingers and some by the thumb. Realizing that people have different sized hands, some manufacturers have made mice that range in sizes from XS to XL and come molded to fit either the right or the left hand. No one device is inherently more ergonomic than another. An input device that fits the tasks, demands and user preferences should be chosen.

5.7.3.3 Additional Equipment (Lights, 10-Key, Phone, etc.)

Additional equipment in the work area may also need to be adjusted to meet task demands. For example, if a task requires telephone answering, the telephone should be close enough that the employee doesn't have to reach far to pick it up. However, if the phone is rarely used, it can be placed further back on the desk. Another common problem with telephones is cradling the phone headset between the shoulder and the ear. This can create awkward head and neck postures. Using a headset is one solution to this problem.

An employee might also do many tasks with one hand. For example, an employee might write, 10-key and answer the phone with the right hand. By switching some of these tasks to the left hand, repetition or duration for the right hand may be reduced.

5.8 Ergoweb Performance Oriented Checklist for Computer (VDT) Workstations

5.8.1 Introduction

Opinions vary with regard to proper computer workstation setup and the consequent body postures assumed by people performing tasks at those workstations. Unfortunately, quick and easy solutions to ergonomic problems are not always available, and a checklist is not able to capture the interactions and complexities of all possible combinations of people, task, equipment and work environment. This is often a source of frustration and certainly a source of misunderstanding with regard to ergonomics. Therefore, Ergoweb has developed this performance-oriented checklist, which intentionally avoids specific dimensional constraints in favor of a principles-oriented approach to the design and setup of a computer workstation.

While using this checklist, remember that designing for ergonomics requires understanding and consideration of the physical and psychological attributes of the person or population of people who will perform the job, the design and arrangement of the workstation and furniture, computer hardware, computer software and other workstation accessories, the tasks required to perform the job, the work environment, including noise and temperature, as well as management and organizational methods and constraints.

The interaction between these general topics is critical and will define the postures, forces and repetitions assumed by the person(s). Further, since a modification in one area may have significant effects in another, no single change can be performed without considering the overall effect. For example, lowering seat height so that someone may comfortably rest his/her feet on the floor may force a stressful upper body posture if the monitor position and table or keyboard height are not adjusted in concert (this is often a good reason to provide a footrest instead).

5.8.1.1 When Using This Checklist

- Ask not whether the person can merely achieve these general goals, but whether the design of the workstation, task and environment interferes with, obstructs or outright inhibits a person from achieving their goals.
- Remember that this checklist is worded for use when reviewing one person and that individual's computer workstation, tasks and working environment. If more than one person must use the same workstation, the checklist should be applied to each individual, and an easily adjustable workstation becomes even more important.
- Remember that there is no "perfect posture," and that a dynamic posture (frequent changes in posture) is a good way to reduce stress and redistribute pressure related to long duration static postures.
- Remember that it is not all-inclusive and may not cover topics important to your specific situation. You may want to adopt certain parts of this checklist for your particular purposes and add your own questions were desired.
- Remember that a good ergonomics approach will improve productivity and quality as well as health and safety.

Ergoweb Performance Orientated VDT Workstation Checklist

Job:	Employee:
Analyst:	Date:
Comments:	

"NO" responses indicate conditions that may be associated with higher risk of illness/injury.

Lower Extremity	**Yes**	**No**
Is the person able to rest the feet comfortably?		
Is the person able to sit with the knees in a comfortable position?		
Is the person free of uncomfortable pressure points, obstructions or other interference in the lower extremities?		
Upper Extremity	**Yes**	**No**
Is the person able to work with the head in a position such that the neck is not stressed by holding the head off balance from the neck and shoulders?		
Is the person able to work with the head facing forward of the plane of the upper body the majority of the time, such that repetitive or long duration head rotation is minimized?		
Is extended reaching minimized, especially where the reaching is held for long durations, is repetitive or requires trunk/torso deviations?		
Is the person able to work with comfortable arm positions or approximately:		
neutral shoulder positions?		
neutral wrist positions?		
comfortable elbow positions that do not force shoulder or elbow positions from approximately neutral?		
Force	**Yes**	**No**
Can all objects that require lifting be lifted forward and close to the body, such that extended reaching and reaching to the side or behind the body is avoided?		
Are all objects that require lifting or moving while seated relatively low in weight or force?		
Static Body Posture	**Yes**	**No**
Do the workstation design and job requirements allow frequent movement and posture changes such that long duration static body positions are avoided?		

Seat Surface	**Yes**	**No**
Is the height adjustable, such that the person is able to set it at an individually comfortable height in relation to the required work activities?		
Is the seat surface of appropriate size, such that it is deep and wide enough to comfortably accommodate the specific person?		
Is the seat slope adjustable, such that the person is able to achieve a comfortable angle, either forward or rearward sloping?		
Is it comfortable and is the front well rounded ("waterfall" front edge), such that the person does not experience excess pressure on the under side of the leg due to the forward edge?		
Overall, is it comfortable to the person who is required to use it?		
Seat Backrest	**Yes**	**No**
Can the person easily adjust its height to provide lumbar support (lower back region)?		
Can the person easily adjust its angle relative to the seat surface?		
Can the person easily adjust it to alter the depth of the seat?		
Overall, is it comfortable to the person who is required to use it?		
Work surface	**Yes**	**No**
With the feet and legs in comfortable positions, can the person achieve a comfortable work surface height?		
Is the width of the work surface appropriate, such that all required task accessories and duties can be located within comfortable reach and viewing distance?		
Is the depth of the work surface appropriate, such that the computer, and keyboard if necessary, can be placed directly forward of the person with the work orientation parallel to the plane of the upper body?		
Is the area under the desk large enough to accommodate the legs and any accessories, such as footrests and arm rests?		
Computer Monitor (VDT)	**Yes**	**No**
Is the person able to easily adjust the relative height of the monitor?		
Is the person able to easily adjust the relative fore-aft distance of the monitor?		
Is the person able to easily adjust the tilt (up/down) angle of the monitor?		
Is the person able to easily adjust the yaw angle (left/right rotation) of the monitor?		
Keyboard	**Yes**	**No**
Is the keyboard detachable from the display unit?		
Is the person able to easily adjust the angle of the keyboard?		
Is the person able to easily adjust keyboard height?		
Is keystroke pressure comfortable to the person?		

Mouse (or other input device)	Yes	No
Is the shape and button activation comfortable and easy to operate for the person?		
Is the person able to operate the mouse without extended, long duration or repetitive reaching?		
Document Holder for Data Entry Tasks	**Yes**	**No**
Is there a special holder or support for the source document?		
Is the person able to easily adjust the document holder height, distance and angle?		
Does the device prevent the document from vibrating?		
Is the device located such that the person is not required to rotate the head/neck extensively or for long durations?		
Support for the Hands and Arms	**Yes**	**No**
Are seat or other armrests available?		
Are armrests adjustable?		
Is keyboard palm support available?		
Support for the Feet	**Yes**	**No**
Is a footrest available if needed?		
Is the person able to easily adjust its height?		
Is the person able to easily adjust its tilt?		
Lighting	**Yes**	**No**
Are lighting levels in the monitor area comfortable to the person?		
Is the monitor screen placed such that light from windows and overhead lighting do not cause glare?		
If not, are reducing methods used?		
Are diffusers, cube louvers or parabolic louvers used on overhead lights where screen glare from that source is a problem?		
Are movable task or desk lights available?		
Do work surfaces have a matte finish to reduce light reflection?		
Do windows have curtains, drapes or blinds available to block light where glare from that source is a problem?		
Temperature	**Yes**	**No**
Is the person comfortable with the ambient temperatures, particularly cooler ambient temperatures?		
Does the person complain that the temperature is too cool?		
Is the person comfortable with the temperatures of any equipment or surfaces he/she must contact?		

Vibration	**Yes**	**No**
Is the monitor free from unwanted motions or oscillations due to vibration?		
Has building vibration been eliminated?		
Noise	**Yes**	**No**
Are sound levels at comfortable levels, allowing conversation and other communications without significant effort or discomfort?		
Ventilation	**Yes**	**No**
Is air circulation sufficient?		
Fatigue Control	**Yes**	**No**
Is the person allowed to take rest pauses or breaks from tasks that require long duration or repetitive postures, forces, keying or mousing activities?		
Is there job rotation or substitution of tasks that require a different type of activity where posture, force and repetition hazards have not been addressed by design?		
Vision	**Yes**	**No**
If the person wears bifocal glasses, is he/she still able to view the computer monitor comfortably, without repeatedly tilting the head back or forward or for long durations?		
Are workers with symptoms of eyestrain, burning sensation in the eyes, blurred vision, irritated eyes or headache examined for vision problems?		
Are the words, images and colors on the computer screen easy to see and read?		
Psychosocial Issues	**Yes**	**No**
Does the person have some involvement and control over the work process?		
Is there good communication between the person and supervisors?		
Has the person been adequately trained?		
Is the software "user-friendly?"		

5.9 References and Additional Information

Ankrum, Dennis R., WorkPlace Ergonomics, Sept./Oct. 1996, p. 10-12.

6. Special Environments

6.1 Introduction

The office workstation does not have to reside at a corporate headquarters. With computers being used in environments as diverse as a home business and law enforcement, the office workstation may be the home kitchen table or the passenger seat in a squad car. The same risk factors of awkward posture, force, repetition, duration and contact stress may be present in these non-traditional environments. While the solution may need to be more creative, the principles of finding a solution do not change.

Awkward postures, such as reaching for the mouse, or static postures such as sitting in one position for a long time, can also be risk factors in the home or mobile office. If the duration is significant, the same techniques of moving the equipment closer to the user or introducing posture changes apply.

6.2 Laptops and Mobile Computing

Reprinted from Ergonomics Today™. Used with permission.

Ergonannies Warn Mobile Computing Ads Could Be a Bad Influence

April 16, 2003

Computer users in Australia are being advised that a recent wave of wireless networking ads could be a bad influence on the way consumers interact with their mobile computers.

In an article in the Sydney Morning Herald, ergonomists, physiotherapists and even lawyers denounced the benefits of wireless networking, warning that the televised images associated with a recent push to go portable include images of people working while lounging with their computers - positions that could prove to be unhealthy.

"Advertising often portrays a device user in all sorts of compromising positions. This may encourage the workforce to adopt similar work habits," said Sydney ergonomist Grant Richter. "Lounging is a key visual when selling these products to make work seem like a holiday and hence attract the buyer into thinking that a posture or habit is safe long-term, until they experience some form of pain or discomfort which they find difficult to relate to a cause that may be obvious to an ergonomist or physiotherapist," Richter continued.

Anna Lee, national chairwoman of the Australian Physiotherapists Association, mimicked Richter's concerns, also offering that a recent Intel Centrino ad airing in Australia showing a woman using a laptop computer in an airport lounge was "irresponsible."

Lee noted that in the Intel ad, the computer user has bad posture, displays poor use of the wrists and fingers in keying, uses only one hand to key, has glare on the monitor from nearby windows, and, overall, is encouraging prolonged use of computers while traveling, which may increase the risk of deep vein thrombosis (DVT).

According to the *Sydney Morning Herald*, the Intel ads portray on-the-go workers casually conducting business through wireless networking hotspots while using laptop computers at airports, coffee shops and other unconventional locations. An Intel representative interviewed said the hotspots are not intended to replace traditional office working environments.

Source: Sydney Morning Herald

Publisher's Note:

Ergoweb's Dr. Peter Budnick offers another perspective. "Granted, I've never seen the ads that spurred these comments, but I will point out that ergonomics is not really about telling people how they should sit. I think the bigger picture is that the wireless industry is developing technology that has the potential to increase human performance and comfort, which is the essence of ergonomics. If wireless technologies will give us more flexibility in when and where we conduct our work, I'd find that an improvement. Rather than point fingers at these innovative companies for supposedly encouraging bad habits, I'd rather appeal to them to further embrace ergonomics, which is a user-centered design approach that will improve the usability of their products. Better usability means better customer experiences, which translates to better market penetration, which translates to success. So, the story here is not about 'ergonannies' scolding people for bad work habits, it's about working with innovative companies to ensure that the user remains a central design perspective – for their own benefit. Ergonomic designs elicit good work habits."

– Jeanie Croasmun

Reprinted from Ergonomics Today™. Used with permission.

Laptops Bring More than Technology to Squad Cars

March 22, 2002

Six years ago, Omaha police departments began a $2 million effort to bring technology to the fingertips of officers. What they didn't foresee was poor ergonomics and possible safety hazards.

The plan sounds great; outfit squad cars with laptops so that officers will be able to run background searches, see mug shots and perform other high-tech functions from their cars, which essentially serve as their offices. What has actually occurred are complaints of awkward postures, back pain, possible public safety concerns, and hospital visits for 58 officers who have struck their heads on relocated gun racks!

In February, the Police Department's Safety Committee flagged more than a half-dozen such issues. According to the city's own safety inspector, the laptops have been installed in a way that creates a "future ergonomic nightmare".

Another big concern for officers is that the laptop's location may take their attention away from suspects and cars during traffic stops. The laptops are placed low to avoid potential airbag inflation. The monitor focuses an officer's attention toward the floor of the passenger side as opposed to out the windshield.

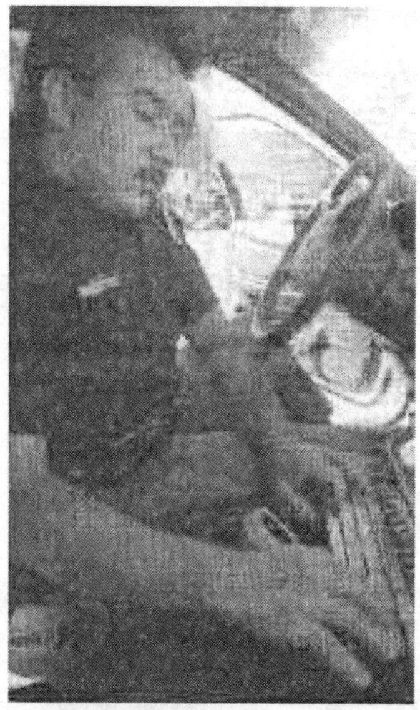

Other ergonomic concerns include:

- Emergency controls which were moved back about 16 inches with the introduction of the laptops. This puts them out of sight and requires shorter officers to bend their arms in a Z-shape.
- Officers have to twist their torsos 30 to 45 degrees and bend forward to use the computers.
- Glare and brightness may prevent the screens from being seen on sunny days.

The Safety Committee did recommend the type of computer that might be used instead. One solution might be a "mobile" computer that has the monitor mounted on the dashboard and a detached keyboard. This saves on space, keeps officers' eyes higher and allows them to move the keyboard. Since intensive typing is not usually required, it was felt the possible misalignment of monitor and keyboard would not be a significant concern.

The project to outfit cars with laptops is currently on hold until these issues are resolved. Half of the planned 121 computers have been purchased.

- Rachel Michael, M.Sc., AEP

Reprinted from Ergonomics Today™. Used with permission.

Taking the Oxymoron Out of Ergonomic Laptops

September 19, 2003

For years, regular users of laptops have known that anything labeled as an ergonomic laptop is either a typo or an oxymoron. Portable and convenient, sure, but the notebook computer's ability to cause pain from tiny keyboards, attached screens, heavy weight and an unfortunate locale for a high operating temperature (see "Got a Hunk of Burning Love? Could Be Coming From Your Laptop", *Ergonomics Today,* November 27, 2002) have made them anything but ergonomically friendly. IBM, however, is trying to change that.

Unveiling two prototypes for the next generation of laptop computers at the TechXNY show in New York, IBM's new designs seek to eliminate user complaints about all portable computers, regardless of manufacturer – they just aren't comfortable for regular use.

In the prototype modeled after the ThinkPad T40, IBM incorporates a hinged display that can be raised up to three inches above the keyboard. The keyboard itself can slide forward and tilt as well. In the second prototype, modeled after desktop computers, the monitor can be raised and lowered, but the keyboard is detachable and wireless. Currently, both are just models; no production date has been announced.

Earlier this year, IBM made another modification to its ThinkPad R40 by offering users a choice of TrackPoints - the red eraser-like pointing device in the center of its keyboard. Users can change their TrackPoint from the original nub to a softer version of the same device or to an all-new style that features a cratered indentation instead of a protruding nub.

Source: PC World

- Jeanie Croasmun

Reprinted from Ergonomics Today™. Used with permission.

Got a Hunk of Burning Love? Could Be Coming From Your Laptop

November 27, 2002

Taking business on the road with you can have its hazards as a scientist in London will attest after he ended up with a burned penis from using his laptop computer. According to *CNN.com*, the 50-year-old unnamed scientist reported a "burning sensation" after using his laptop computer situated on his lap for about an hour while writing a report.

The scientist noticed irritation and redness, but until he was examined by a doctor the following day, he didn't realize the scope of the damage done by his computer.

"The ventral part of his scrotal skin had turned red, and there was a blister with a diameter of about two centimetres (0.8 inches)," wrote Claes-Gorn Ostenson, of the Karolinska Institute in Sweden, in a letter published by *The Lancet* and reported by *CNN.com*. While the wound eventually became infected and crusted, subsequent reports on the unidentified scientist's condition deemed him healing rapidly.

Laptop computers have inherent ergonomics risks associated with their small keyboards, awkward positioning of the user in relationship to the attached screen and keyboard, and the laptop's heavy carry-on load. During normal use, however, burned penises aren't often considered a potential side effect. Even so, Ostenson stated that the computer's manual did warn against using the laptop computer while naked, but the scientist claimed to have been fully clothed at the time of the incident.

Source: CNN.com

- Jeanie Croasmun

6.3 Telecommuting and Home Offices

Reprinted from Ergonomics Today™. Used with permission.

Stress-less Homework Anyone?

October 18, 2002

If taking the work home with you seems like a stressful idea, think again. Recent studies show that it can mean exactly the opposite.

According to an European Union survey of telecommuters at BT, more than 90 percent of the respondents said they experienced less stress and greater productivity because of telecommuting. The survey, reported on *ZD Net UK*, asked 2000 of BT's telecommuting staff for their opinions, both good and bad, on telecommuting. In addition to a less-stress environment, other benefits the respondents listed included the ability to multitask, the lack of commuting time and the flexibility of choosing when to work. Drawbacks were isolation, increased working hours and lack of visibility by higher management.

It is estimated that 20 million people in Europe telecommute at least one day each week, accounting for 7.4 percent of the work force. In 2000, an estimated 16.5 million workers in the United States also regularly telecommuted, although more recent studies suggest that the number could now be upwards of 28 million people. For those ever-present skeptics who claim teleworkers don't get anything done, a study by the International Telework Association and Council reported that nearly three-quarters of all teleworkers said once they began teleworking, their productivity *increased*, while only 6.5 percent reported a decrease in productivity.

But is stress alone enough reason to trade in the pumps for slippers and the morning commute for a programmable coffee maker? According to Ergoweb President and CEO, Peter Budnick, the answer might be no. "Home offices are often thrown together with whatever desk, chair, and other furniture that's available - and usually cheap. If we do go out and buy something, we usually forego adjustability and other options that would make the furniture more flexible for our different body sizes, task needs and individual preferences. We select the $150 products because we can afford them, not because they best fit our needs," Budnick says.

Yet there is hope. Budnick suggests following in-office ergonomics principals and applying them to the home front. "There are a lot of things that can be done without spending a lot, or even spending a dime," he says. "Heights [monitors, desks] can be modified using wood blocks or books. Footrests, if desired, can be fashioned out of boxes, phone books, etc. It may not be pretty, but it doesn't have to be expensive. The primary goal is to reduce awkward postures like reaching too far for your mouse, reaching too high for your keyboard, working with your monitor to the side so that you have your head turned all day, every day." The rule, says Budnick is to listen to your body. "Pain and discomfort are warning signs that something is wrong with either your work area set-up or your work habits. You aren't performing at your best when things are wrong," says Budnick.

But there's little recourse for the stubborn home worker who insists upon setting up shop with a laptop on the kitchen table, and who then finds him or herself with a work-related injury. "At home, the burden is on the worker, not the employer," says Budnick. While threats of regulating home offices, thus making the employer responsible for what goes on behind the employee's doors, have reared their pesky heads in the past, none of the regulation attempts have been successful.

Sources: ZD Net UK, International Telework Association and Council

– Jeanie Croasmun

Reprinted from Ergonomics Today™. Used with permission.

Ergonomics and the Telecommuter

June 18, 2003

Over 28 million Americans went to work today without ever leaving their homes, courtesy of telecommuting. And each day, that number is increasing as employers are warming up to the idea that employees actually can work from home.

In a 2002 survey, AT&T, a vast proponent of teleworking, was found to be saving an estimated $25 million each year on office space in addition to over $65 million annually in improved productivity thanks to its home-based workers.

But as the number of teleworkers increases, the question that arises is how ergonomic is the home environment? Desks crammed into corners, sharing spaces with guest rooms, laptop computers on dining tables and chairs hauled down from someone's attic can all compromise home office ergonomics.

At one point, OSHA tried to intervene. In late 1999, the agency issued an advisory letter regarding telecommuters that suggested employers be responsible for ergonomics at their workers' in-home offices. The recommendations included periodic inspections by employers, a concept that didn't bode well with either employer or employee. Subsequently, the letter was withdrawn and telecommuters were given back the responsibility of making their homes ergonomically sound business environments. With offices that double as kitchens and basements, that's no easy feat.

"The biggest challenge is the space available for a home office," says home office furniture designer Jack Kelly. "The scale of the furniture is very important. I've talked to interior designers who have converted living rooms into home offices. Also, shared spaces are another consideration," Kelley says. People have to be able to live in their houses, says Kelley, and very few people purchase a house with the notion that they may need to secure an extra bedroom for their employer. Plus, ergonomics are rarely at the top of the manufacturers' to-do lists.

It's not just the manufacturer, however, who takes the blame for the somewhat compromised ergonomic conditions of the home-based office. Employees tend to look for a couple of things, says Kelley; they either want inexpensive furnishings which could mean hand-me-downs from other parts of the home or they want styles that blend in with their decor – something that traditional office furniture has never been good at.

Some employers take the matter of ergonomics into their own hands. At Allina Metro Hospitals in Minnesota, for example, over 50 percent of the transcriptionists telecommute. According to Allina representative Barbara Lietz, the telecommuting program, which was started over 11 years ago, includes an on-site in-home evaluation. Plus, all telecommuters are provided an ergonomic chair and a computer and resource materials. "We are concerned for the well-being of our employees ... medical transcription is a sedentary job and it is essential to implement ergonomic accommodations," says Lietz. But not all employers take the extra steps to ensure that telecommuting employees' workspaces are ergonomically sound.

Designer Kelley sees hope on the horizon for the ergonomics of telecommuting, noting design trends that incorporate appropriate lighting conditions into the furniture, office furnishings that are being scaled down to fit the home environment, and concepts like workstation armoires that function well as workstations when open but have the appearance of a well-designed hutch when closed. Kelley, who has been designing office furniture for decades, is also seeing more homes built with dedicated office spaces, indicating that the cramped multi-purpose room could eventually become a thing of the past. For the estimated 28 million Americans who never left their home today but still managed to punch the time clock, that all sounds like good ergonomic news.

Editor's Note:

For more information on setting up a home office with ergonomics in mind, see the June 2003 issue of The Ergonomics Report.

Source: Data courtesy of International Telework Association and Council

- Jeanie Croasmun

Reprinted from The Ergonomics Report™, Vol. 2, No. 6, June 2003. Used with permission.

Who Takes Care of Ergonomics In the Home Office?

To the people who spend their days in an office, the concept of telecommuting might sound glamorous. Stroll out of bed, pad down to the kitchen for a cup of coffee, and make business contacts while wearing fuzzy slippers - a dream-come-true work environment.

But for the home worker, that "dream" can also equal an ergonomics nightmare. Aside from the comfortable slippers, that same job that invades the house requires space and equipment and a host of other accommodations to work with the employee and fit into the employee's home life. But who is responsible for supplying these?

That depends on whom you ask.

In 2001, 28 million Americans telecommuted, either on a part- or full-time basis. While some of them worked on the road or in satellite offices, most of these telecommuters did jobs from their home-based offices. Desk, lamp, computer, chair; the equipment is all there, but in what form and how is it being used?

On occasion, Carolyn Lundberg, president of ACE Ergonomics in Newport Beach, California, is called upon to review a home office environment. Her prognosis? Not so good. On a recent trip to see a worker's setup, Lundberg was escorted past the kitchen, bedrooms, living room, study, through a sea of toys, and finally to the worker's home office. "Someone else had built this office and it would have been fine if you weren't going to use a computer there," Lundberg said. But the worker was working on a built-in shallow desk with shelves above it, leaving little room for his laptop computer. Behind the desk was a 30 inch table – a perfect place for the worker to spread out with his computer and his work, to have elbow room and the ability to reposition himself and move around, but the children's computer was there instead.

Lundberg sees a lot of this. "It's almost always a hodgepodge," Lundberg says of the furnishings. Discarded writing desk, folding chairs, space shared with toys or books, or even setting up shop at the dining room table. Given the same setup at their employer's office, Lundberg says, these telecommuters would probably be complaining.

In late 1999, OSHA issued an advisory letter regarding telecommuters that suggested employers be responsible for the home office environment as well. Included in the letter were recommendations for periodic reviews of the home-office environment by the employer. The ensuing flack from business owners who felt it was inappropriate for them to be held accountable for the employee's private home, and telecommuters who had no desire to see their bosses come traipsing through their spare bedrooms, caused the agency to quickly withdraw the advisory letter.

So what drives the telecommuter's office? According to Jack Kelley, a home office furniture designer, that depends on the telecommuter.

"The biggest challenge is the space available for a home office. The scale of the furniture is very important," says Kelley. "I've talked to interior designers who have converted living rooms into home offices. Also, shared spaces are another consideration," Kelley says. People have to be able to live in their houses. Very few people purchase a house with the notion that they may need to secure an extra bedroom for their employer. But what about ergonomics? Kelley says it's rarely at the top of the home office furniture manufacturers' lists.

As for what drives home office furniture design, Kelley says that "a lot of people focus on inexpensive furniture." He also notes that furniture that blends into an existing decor or that can be readily made to look like something other than an office desk, such as a closed armoire, is also in demand.

"One of the more important parts is space to move around in," says Kelley. Corporate offices tend to be larger. Over the years, Kelley has seen trends where people were converting spare bedrooms into offices; now more people have a dedicated office space in their house. But still, Kelley's job is to make the furniture practical, appealing and in a scaled-down size that fits a house. Hopefully it fits the worker well, too.

Still, the matter at hand is who becomes responsible for the ergonomics of the home office. Ergonomist Lundberg admits that employees usually point the finger at employers and employers tend to point it back at the employees. Other people will say it depends on who is requesting that the employee work at home. If it's a convenience for the employee, the employee should be responsible. If it's a convenience for the employer who lacks appropriate office space or who needs a representative in a satellite location, then maybe the employer should pay up. No employee would be required to buy his or her own desk or chair while working at the corporate office. But how many employees want the standard-issue office cubicle and ergonomic chair sitting in their living rooms?

- *Jeanie Croasmun*

Reprinted from Ergonomics Today™. Used with permission.

The Office Workstation at Home

May 7, 2001

Have a home office? You are not alone. More and more people are finding the convenience of a home office a must. It is used for keeping up with friends and family, organizing personal finance and for occupational work.

What many don't know is that home offices can present some of the same risk factors for the development of musculoskeletal disorders (MSDs) as your company office.

Home offices do have many benefits. You are more able to control your speed of work, when you can take breaks and duration of any one task. You may not suffer from extra stress related to personal relations at work, or you may not be exposed to as much noise.

While these are all good things associated with the home office, there are also some negative trends. The computer workstation is one of the top concerns. Many people use a coffee table, the kitchen counter or other objects to serve as the workstation. The problem with this is that these surfaces were not designed to be used with a computer, making the user more likely to assume awkward postures while working. A common chair for these home workstations is the old dining set chair that you only pull out when grandma comes for dinner. Again this chair was not designed for computer use. It is not adjustable and does not provide the support needed for extended periods of sitting.

If the computer workstation also doubles as the kids' game/homework station, a proper fit is even more important. If it is not possible for the child to get their own workstation designed with their dimensions, some extra adjustments should be made. Pillows or phone books or whatever will help raise the child will reduce the amount of reach they have in their arms. Workstations for children are available on the market and can be found at www.ergobuyer.com.

Some manufacturer also produce keyboards and other input devices designed specifically for children's hand sizes.

The following report released by the American Furniture Manufacturers Association (AFMA) also has some tips for your home office. For more information about this, visit the AFMA.

AFMA Report

Nearly one-third of American households have a home office, according to a study conducted by the American Furniture Manufacturers Association (AFMA). Whether it's located in the den, a spare bedroom or even exists as part of another room, such as the kitchen or master bedroom, the home office has become an important part of our culture today, and its popularity and prevalence are only expected to increase. Fueled largely by trends in technology and telecommuting (nearly half of all U.S. households now have a personal computer), the home office is clearly here to stay.

"Consumers don't have to worry that home office furniture is cold or commercial in its look, or that one size fits all," says Jackie Hirschhaut, AFMA vice president. "The wide variety of home office furniture and accessories available today can reflect your lifestyle and personality, and even complement other rooms in your home."

Making Room for Work and Play

According to the AFMA survey, of those with home offices, nearly 40 percent use their office primarily for entertainment, to surf the Internet, send e-mail or shop; 30 percent use it for work; and 22 percent use it to organize their home and pay bills. In addition, 55 percent of those surveyed spend more than five hours per week in their home office.

Considering the extensive variety of everyday uses for a home office, it's difficult to understand why some consumers deny themselves a home office. "People may not recognize when they're ready for a home office and may not realize how home office space can provide so many benefits – whether it's a room of its own or a part of another room," Hirschhaut says. "And there are so many furnishings options in all price ranges so anyone can – and should – enjoy office space wherever it works best in their house."

Indeed, nearly half of today's home offices currently share space with another room. The master bedroom pulls double duty most often, followed closely by the family room. "People can still pay bills, send e-mail or work in the family room if that's the most convenient place for them," says Hirschhaut. "They don't have to use the coffee table as a workstation anymore."

Put the "Fun" in Functionality

According to the AFMA survey, the desk was named the most practical piece of home office furniture. On consumer wish lists for home office furniture, the chair ranked highest. "Considering the popularity of ergonomics as well as the unlimited number of colors and designs available, this is no surprise," Hirschhaut says. "There is so much more variety than there was 10 years ago. Why settle for your father's office chair?"

In addition to the variety of chairs, today's workstations, desks, filing cabinets and lamps come in all shapes and sizes. "Consumers have many more options today. For those just starting a home office or even for those looking to add versatility to their current space, just start with a desk and add on pieces that you need," says Hirschhaut.

All the Comforts of Home

Home offices offered by manufacturers can meet the needs and specifications of every individual or business. The common thread is that they can be personalized to reflect the needs of families or individuals who use them, as well as create an atmosphere that is both functional and fashionable. "One person might prefer a more elegant or traditional workstation, while another may want today's high tech colors and sleek designs. Fortunately, there is a piece that suits every taste and style," says Hirschhaut.

"Just like any other room in your home, your home office should reflect who you are, what you like and what's comfortable for you," Hirschhaut concluded.

Quiz: Are You Ready for a Home Office?

Do the bills in your household pile up on the counter? Does your kitchen table double as a computer desk? Take this quiz to determine if you're a candidate for a home office.

- Does your home computer make its home on your kitchen table?

- Do your dresser drawers double as your filing cabinet?
- Would you or your children benefit from an organized, well-lit workspace?
- Do you have some unused space in your home that could serve a more functional purpose?

If you answered "yes" to any of these questions, you may be ready for a home office.

- Rachel Michael, M.Sc., AEP